Postwar Soviet Politics

The Politics of Soviet Agriculture, 1960–1970

Postwar Soviet Politics

THE FALL OF ZHDANOV AND THE
DEFEAT OF MODERATION, 1946–53

WERNER G. HAHN

Cornell University Press

ITHACA AND LONDON

International Standard Book Number 0-8014-1410-5
Library of Congress Catalog Card Number 81-15234
Printed in the United States of America
*Librarians: Library of Congress cataloging information
appears on the last page of the book.*

For Marina

Contents

Contents

Preface

The generally accepted interpretation of postwar Soviet politics holds that in 1946, with the initiation of a harsh ideological crackdown and the onset of cold-war tension, tight Stalinist controls were quickly reimposed and that Stalin's top deputy, Andrey Aleksandrovich Zhdanov, was the chief executor of this new hard line. Indeed, the shrill ideological campaigns, violent purges, outbursts of exaggerated Russian chauvinism, and strident cold-war rhetoric that marked the period from 1946 through 1949 have come to be known as the *Zhdanovshchina*. My research suggests, however, that this period, rather than being devoid of intellectual exchange and political debate, witnessed an intense competition between relative "moderates" and "extremists" in the Soviet establishment and that Zhdanov, surprisingly enough, stood on the "moderate" side and was in fact the chief victim of the triumph of extremism. In reexamining the Soviet press of this period and studying the extensive source material that has appeared more recently, I find much evidence that divergent viewpoints, rather than monolithic orthodoxy, characterized Soviet officialdom after the war, and that the dominant political forces in 1946 — Zhdanov and his followers — were dramatically overturned by 1949, as part of a historic defeat of moderate elements in the Soviet political establishment.

During World War II the police terror that had ravaged all segments of Soviet society was eased in deference to the war effort, and the concomitant relaxation of political controls and ideological standards provided more room for creative pursuits. By the end of the war, the scholars who dominated such fields as philosophy, economics, and natural sciences expected that both this relaxation and the wartime cooperation with the West would be continued, and they spoke out accordingly. The concept of a single world science, rather than the notion of separate "capitalist" and "socialist" sciences, was openly promoted by natural scientists and philosophers, and Western ideas were respected in fields such as biology. The lively debates that occurred in economics indicate that there was still room for a certain amount of objectivity and pragmatism in that field, as well.

Statements made in early 1946 by such political leaders as Zhdanov, and even Stalin himself, suggest that they too expected a period of relative relaxation at home and abroad. The fact that Zhdanov initiated an ideological crackdown in 1946 does not alter this picture, first, because the reimposition of controls appears to have been more a function of political maneuvering than an expression of ideological principle, and second, because Zhdanov did not extend this crackdown in literature and culture to parallel fields of intellectual activity. Moderates remained prominent in politics and ideology until 1948–49, when Stalin stripped Zhdanov and his supporters of power and backed the most extreme chauvinists and dogmatists in their attempts to purge the social and natural sciences of unorthodox thought.

Throughout the immediate postwar period, these moderates formed the establishment in social and natural sciences and even in the ideological field. Some could be characterized as "reformist" or "liberal" because they advocated changes that would have softened the harsher features of the system. Others simply expressed relatively objective views, tolerated divergent ideas, or resisted the imposition of uniformity, and hence can be described as "moderates" because of their contrast with the extremists who later became so predominant.

While some of these moderates later changed in an often vain attempt to save their positions, others held firm. The clearest example of the latter is the philosopher Bonifatiy Mikhaylovich Kedrov, who actively promoted freer inquiry in philosophy and fought Lysenko's attempts to shut out Western science and impose his dogma on Soviet biology. Evidence of Kedrov's "liberalism" can be found not only in the accusations made by his hard-line critics—often the main source for such labels—but also in his writings, where such views are clearly expressed. Moreover, Kedrov continued to voice such "liberal" views well into 1948, and only the full triumph of dogmatism suppressed his voice.

The opposing camp—the "dogmatists"—was motivated both by opportunism and by principle. Since relative "moderates" controlled most of the key positions in social and natural sciences, some dogmatists may have taken a superorthodox stance simply in order to seize posts held by the establishment figures. But the best known of them—philosophers Pavel Yudin and Mark Mitin and biologist Trofim Lysenko—had long ago earned their reputation as "dogmatists," both by their habitual insistence on one truth and by their intolerance of other, more creative views. These men resented the fact that the ideological and scientific community rejected their views while Zhdanov's political machine thwarted their ambitions, and they eagerly took advantage of Stalin's shift toward dogmatism to pillory their foes.

This postwar defeat of moderate elements was a historic turn in Soviet political life, and much of the subsequent course of Soviet political and intellectual development has turned on the efforts of scientists and scholars to undo the damage wrought in 1948–49. After Stalin's death, and especially after Khrushchev's denunciation of Stalin, moderates and liberals (both in politics and in the social and natural sciences) struggled to reduce oppressive party control and establish more room for objective scientific inquiry. They not only challenged the domination of the dogmatists and reasserted the right to adopt useful Western scientific and scholarly ideas, but even sought to limit party interference in science and philosophy. After Khrushchev's ouster, however, the increasing conservatism in political leadership eventually led to moves to block further loosening of controls and in fact reversed some of the earlier progress. The fact that some of those who had helped impose dogmatism in 1948–49—for example, Central Committee Secretary Mikhail Andreyevich Suslov—are still in power and continue to resist relaxation has also played a role in this continuing frustration of tendencies toward reform.

The role of Zhdanov in these developments appears to have been quite different from that portrayed in most Western histories and also from that acknowledged by intellectuals and politicians in the Soviet Union itself. Evidence suggests that despite his image as the champion of militant orthodoxy, it was not Zhdanov who led the ideological dogmatists and Russian chauvinists and brought about the defeat of moderates. Rather, it appears that Zhdanov's ideological campaign was motivated more by essentially political concerns: the circumstances of its origin, the method of its prosecution, and the identity of its executors mark it as part of Zhdanov's broader campaign to defeat his top party rival, Georgiy Maksimilianovich Malenkov, and to reestablish the primacy of the ideological apparatus—Zhdanov's main base of power—in party life. The position of ideological workers had been badly eroded during the war, when practical needs raised the importance of economic administrators and military leaders, hence enhancing the power of Malenkov, the cadre specialist and industry supervisor.

In the course of restoring the primacy of the ideologists, Zhdanov made vicious assaults on those who neglected ideology, especially during his highly visible campaign in cultural affairs, a policy that earned him his reputation as the regime's leading hard-liner. In fact, Zhdanov was neither as dogmatic nor as chauvinistic as his reputation suggests. He did not seek to impose narrow ideological constraints in fields such as philosophy and natural science, but in fact resisted dogmatism and encouraged creativity in these disciplines. He did not appear to believe that ideology had to dictate an answer for everything, and, cultivating an image as the Politburo's "intellectual," he encouraged some philosophers and biologists who held unorthodox views. Moreover, in late 1947, as the cold war intensified and the dogmatists found increasing favor with Stalin, Zhdanov began to lose power and his apparent protégés began to be accused of ideological softness and internationalism. What is loosely called the *Zhdanovshchina* actually climaxed after Zhdanov's death, with Lysenko's ruthless purge of science, Stalin's drives against cosmopolitanism and the Jews, and the use of punishments—including execution—more extreme than those that had occurred during Zhdanov's heyday.

Of course Zhdanov, who had been involved in Stalin's bloody purges in the 1930s, was hardly a "liberal," even by Soviet terms. In fact, in 1946 he cynically sought to beat his enemies by campaigning against ideological softness and by enunciating crude, doctrinaire attacks on culture. Although Zhdanov was initially successful with this tactic, his rivals eventually used a similar approach to defeat him, taking advantage of his apparent patronage of dangerous "liberals" and unpatriotic internationalists to maneuver him into a vulnerable position. We can tentatively label Zhdanov a "moderate" or patron of moderates simply because his positions, at least initially, were less extreme than those that became official in 1948 and 1949. Zhdanov's views may have hardened later, but the facts that he lost influence as the atmosphere grew more dogmatic and that his earlier views appear to have been held against him indicate that, willy-nilly, he wound up in a "moderate" position. To some extent, Zhdanov's "moderation" may have resulted simply from a failure to foresee how far Stalin would go in sanctioning dogmatism and repression. Such a misjudgment by Zhdanov is certainly conceivable, for Stalin's extremism resulted, at least in part, from his turning against Zhdanov personally and his related decision to encourage the eradication of Zhdanov's faction and to proscribe the "moderate" views that Zhdanov had apparently permitted.

In developing this picture of the immediate postwar period, I have relied heavily on analysis of the activities of sub-Politburo officials and of struggles within the party apparat—tracing the political connections among ideological officials, as well as the arguments and accusations in the debates over philosophy, science, and economics. The period was filled with fast-moving events—a high level of intrigue and turmoil, sharp debates, and sudden reversals of individual political fortune. Close examination of the period's central and local newspapers and journals reveals that there was probably enough information available at the time to solve many of the puzzles, but the collection of extensive materials now available—newly published decrees, obituaries, biographies, and memoirs—has greatly aided in this task.

The first four chapters of this book roughly correspond to individual years, each of them constituting a distinct stage both in the story of Zhdanov's rise and fall and in the contest between moderates and dogmatists. Chapter 1 focuses on Zhdanov and his supporters in 1946, detailing their views and examining their successful defeat of Malenkov. Chapter 2 analyzes the 1947 debates in philosophy, biology, and economics—debates that led to the defeat of the moderates and marked the start of Zhdanov's decline. The final intrigues against Zhdanov in 1948, and his subsequent fall and death are covered in Chapter 3, while Chapter 4 describes the triumph of obscurantism in philosophy and science and the purge of Zhdanov's protégés in 1949.

Chapter 5 turns to Stalin's last years, carrying the story of the struggle among Soviet leaders and its relationship to "moderation" on through Stalin's final intrigues to the triumph of Khrushchev over Malenkov after Stalin's death. Much in the 1950–53 period remains obscure, however, since Stalin's plans for a new purge were aborted by his death and he had carefully hidden his intentions and many of his actions from the outside world and even from his closest colleagues.

In contrast to the other chapters, Chapter 6 examines the career of a single man, the philosopher B. M. Kedrov, in order to focus on the question of whether "liberals" exist in the Soviet establishment. Although the effort to erase the stultifying influence of the late Stalin period has been manifested in many disciplines—economics, historiography, sociology, psychology, and so on—the forefront of this struggle was in philosophy and biology, the fields that were Kedrov's most direct concern. His persistent attempts—both during 1946–49 and in later years—to drive out the dogmatism and party interference that have hampered and discredited Soviet science are perhaps the clearest illustrations of the long-term struggle between "liberals" and "dogmatists."

In addition to the light it sheds on the existence of reformers in the Soviet establishment, the story recounted in this book raises some important questions about the late Stalin period: Was the defeat of moderation unavoidable? How total was Stalin's domination of Soviet politics? What role did Stalin's lieutenants play in the political process? A short conclusion addresses these questions.

Finally, in view of the extreme difficulty of obtaining reliable biographic and organizational information on the postwar Soviet leadership, I include appendixes that summarize all the available press information on membership in the Politburo and Secretariat, as well as on the staffing of leading posts in the government and CC apparat. These data are presented in the form of charts with supporting documentation. Because of the frequent reorganizations of the CC apparat during the 1940s and 1950s, I also include a detailed history of changes in the structure of CC departments.

I wish to thank those who have helped by reading parts of this manuscript and offering suggestions and criticisms, including Paul Goble, John Haskell, Tom Bjorkman, and Marc Zlotnik. In addition, I express appreciation to Setrag Mardirosian for his many criticisms of the manuscript, ideas on overall interpretation, and specific proposals for further work, especially regarding policy positions. I also thank the Russian Institute of Columbia University and George A. Bournoutian for their assistance in making available transcripts and tapes of Khrushchev's recollections.

WERNER G. HAHN

Silver Spring, Maryland

Postwar Soviet Politics

CHAPTER I

Zhdanov's Victory — 1946

At the end of the war, after three years of relative isolation from Moscow while leading the defense of Leningrad, Central Committee Secretary Andrey Aleksandrovich Zhdanov returned to the capital and by late 1946 had established himself as Stalin's clear favorite and heir apparent. Zhdanov's rise came mainly at the expense of another CC secretary, Georgiy Maksimilianovich Malenkov, who had become Stalin's powerful deputy while running the party and government apparatus in Moscow during the war. Zhdanov triumphed by pointing out to Stalin the ideological and administrative laxity that had developed in the party apparat under Malenkov and by launching campaigns to restore the priority of ideology and rectify abuses in party administration, agriculture, and finally, literature and culture.

Malenkov had supervised operations of party organs and selection of party officials during the war, and his power was officially recognized in early 1946 when he was elected a full member of the Politburo and given a status equal to Zhdanov in protocol rankings. Zhdanov reacted to his rival's rise by launching a campaign to convince Stalin that Malenkov had done a poor job in managing the party apparat and selecting cadres. He asserted that under Malenkov, party officials had focused too much on such matters as production and had become lax in matters of ideology, tolerating dangerous "liberal" trends in literature, culture, historiography, and local nationalism. Putting his ideological apparat to work, Zhdanov quickly produced evidence to support these contentions. Stalin, convinced of Malenkov's negligence and of the need for a renewed stress on ideology, removed Malenkov from the Secretariat and gave

Zhdanov and his allies the go-ahead to purge and reorganize the CC apparat and to begin an ideological crackdown. Since this drive was launched in early 1946, the crackdown in literature and culture —the infamous *Zhdanovshchina,* which was publicly announced in late August 1946 —was actually only the culmination, if not the aftermath, of the process.

Despite his leadership of this campaign, Zhdanov appears to have patronized the forces of relative moderation, for he was hostile to the dogmatists in ideology, philosophy, and science, promoting and protecting his own clique in these fields. Moreover, although evidence of the policy positions of Kremlin leaders is usually tenuous, Zhdanov's statements suggest that he took a relatively relaxed view of world affairs and favored a shift away from defense industries to a greater concentration on the satisfaction of consumer needs. Zhdanov's campaign for restoration of the importance of ideology was not a call for the reimposition of dogmatism but a reflection of bureaucratic politics: Zhdanov's ideological apparat had been downgraded during the war and Zhdanov was attempting to restore its influence. Indeed, relative moderates in such disciplines as philosophy, biology, and economics flourished until late 1947 and early 1948. The rise of dogmatism in these fields coincided with the decline in Zhdanov's influence, and the dogmatists triumphed fully only after his demise.

Zhdanov as a Moderate

Despite his reputation as a hard-liner, Zhdanov appears to have been a more moderate influence than Stalin's other top deputies, Foreign Affairs Minister Molotov, CC Secretary Malenkov, and police chief Beriya, at least in 1946–47. His statements during 1946 suggest that he expected relations with the West to remain relatively good, with no danger of a new war. This continued relaxation in the world situation naturally would permit an increase in the production of consumer goods and an improvement of living standards, and Zhdanov appeared to foresee such a shift of emphasis in economic policy. In contrast, Malenkov, Beriya, and Molotov stressed international dangers and the consequent need for stronger defense. Even though Zhdanov called for a greater emphasis on ideology, he resisted the dogmatists who sought to impose narrow ideological constraints in philosophy and science and encouraged more creativity in these fields.

Zhdanov's apparent moderation probably reflected not a basic difference with Stalin but rather Zhdanov's own perception of the world situation and his reading of Stalin's current attitude. Stalin's deputies would hardly dare differ with Stalin on policy issues, given the leader's inclination to strike down anyone showing independent attitudes. But Stalin's views were not always clear and they could change with time or circumstance, sometimes catching even his most cautious deputies out of step. Zhdanov presumably expected Stalin to continue more lenient policies in several areas, even though he himself played up to Stalin's anger over relaxation in party administration and ideology. When Stalin increasingly turned to hard-line positions in foreign policy, philosophy, science, and economics, Zhdanov found himself in an exposed position. Stalin's hostility to Zhdanov in 1948 appeared to originate both in personal jealousy over Zhdanov's earlier prominence and in displeasure with Zhdanov's supposed "softness" or "liberalism."

Much of the direct evidence that Zhdanov expected and favored more relaxation at home and abroad is to be found in his February 1946 speech, delivered in his campaign for election to the Supreme Soviet and printed in the 8 February 1946 *Pravda*. Zhdanov stressed the reorientation of the economy to peacetime production and the satisfaction of the long-suffering public's need for consumer goods, while making no mention of any threat of international tension or renewed armed conflict. He flatly declared that "we have entered a period of peaceful development" and industry is being "transferred to production of peacetime products." He noted that the Soviet people have long borne great sacrifices and now "legitimately demand" that living conditions be improved rapidly. Hence, he argued, the country must tackle the expansion of consumer goods production and improvement of living conditions with "the same Bolshevik enthusiasm and passion" with which it tackled the war effort.

Although such statements may be interpreted as simply reflecting the optimistic mood of the period, Zhdanov's were notably stronger and more committed than those made by the other leaders during the same campaign. In especially sharp contrast, Malenkov virtually ignored the prospects for peaceful development and focused on the international danger and the consequent need to continue the concentration on building military strength and heavy industry. Repeatedly reminding his listeners that only the strong are respected in world affairs, Malenkov concluded that we "must first of all strengthen our might, our force, strengthen our socialist state, strengthen our glorious Red Army and Navy."[1] He warned anyone who might think of organizing new military attacks that the Soviet Union would not be frightened. He stated that the 1946–50 five-year plan, about to be unveiled, would "further develop our heavy industry, the main base of industrialization," and he mentioned consumer goods only in passing (*Pravda*, 8 February 1946). Beriya declared that "it would be mistaken to think that now the need for further strengthening of the military-economic might of the Soviet state had disappeared," and he warned that the workers would still have to suffer privations to keep up a strong defense. He maintained that "comrade Stalin has never lost sight of the fact that sooner or later the Soviet state will have to withstand the onslaught of hostile imperialist states" and cited a Stalin statement that the "main question of the present day" is the "threat of a new imperialist war," specifically, "the real threat of a new war in general and a war against the USSR in particular" (*Pravda*, 6 February 1946). Molotov also underscored international danger and argued that the government would do everything to make its armed forces second to none; he barely mentioned consumer goods (*Pravda*, 7 February 1946). Even Foreign Trade Commissar Mikoyan, long associated with the field of consumer goods production, hardly mentioned them in his speech (*Pravda*, 7 February 1946), and Gosplan Chairman Voznesenskiy, who was shortly to present the new five-year plan, called for continued strengthening of the armed forces but ignored the subject of consumer goods (*Pravda*, 9 February 1946).

Stalin's speech, delivered two days after those of Zhdanov and Malenkov, appeared more in line with Zhdanov's remarks, for it was optimistic about peaceful development and made no mention of new international dangers or the need to focus on heavy industry or defense in the upcoming five-year plan. Instead, Stalin promised to end rationing soon, to give "special attention to expansion of consumer goods production," and to improve living conditions "by consistently lowering prices on all goods" (*Pravda*, 10 February 1946). Apparently Stalin at this point expected relatively smooth

[1] This concern for defense was also notable in Malenkov's next public speech, delivered at the founding of the Cominform in September 1947. On that occasion, he stressed that further strengthening of the armed forces with the latest equipment was necessary to avert new aggression against the Soviet Union (*Pravda*, 9 December 1947).

international relations and hence anticipated that it would be possible to satisfy more of the public's needs.

The five-year plan (1946–50), introduced at the March 1946 session of the Supreme Soviet by Deputy Premier and Gosplan Chairman Voznesenskiy (*Pravda*, 16 March 1946), appeared to reflect Stalin's optimism. Though not specifying the comparative growth rates for industrial "Group A" (heavy industry) and "Group B" (light industry), Voznesenskiy declared that growth in production of the means of production would be "somewhat" higher than growth of consumer goods production—suggesting that the gap between the growth rates of heavy and light industry might not be as great as in most five-year plans. In fact, the output of light industry grew much faster in 1946 than that of heavy industry, largely because the changeover to peacetime production was much more disruptive to the heavy industrial sector. From 1947 on, heavy industry again vastly outstripped consumer goods in growth.[2] Defense spending apparently was down in 1946 also, and seems to have begun an upturn only later in 1947.[3]

Zhdanov's 6 November 1946 speech on the anniversary of the October Revolution somewhat bolstered the evidence that he favored and expected more relaxation abroad, as well as at home—although here his words are probably more representative of the leadership's collective views than had been the case in his February campaign speech. In the anniversary speech, Zhdanov declared that the goals of lowering prices of goods and ending rationing—both promised by Stalin in February—required "significant expansion in

[2] Figures for 1945–50 industrial growth cited in *Narodnoye khozyaystvo SSSR v 1958*, p. 138, show that total industrial production in 1946 was 17 percent below that of 1945 and that Group "A" had dropped sharply (to only 73 percent of 1945) while "B" actually rose somewhat (113 percent of 1945). "A" rebounded in 1947 (to 22 percent over 1946, compared to a 21 percent rise for "B") and thereafter remained above "B." A. Zalkind in *Planovoye khozyaystvo* (no. 4, 1976, p. 62) declared that actual growth of "B" during 1946–50 was higher than for "A," averaging out to 15.7 percent a year versus only 12.8 percent for "A." He explained that this was because the disruption caused by changeover to peacetime production in 1946 had fallen "wholly on Group A" and that if 1946 is excluded, "A" grew an average of 25.6 percent a year in 1947–50 and "B" by only 16.3 percent.

[3] During 1946 the armed forces were cut from a wartime high of 11 million to 3 million, according to Alexander Werth (*Russia: The Post-War Years* [London, 1971], p. 333). But, Werth stated, it was well known that in the second half of 1947 Soviet military manpower began expanding again, to over 5 million. A statement by Khrushchev in his January 1960 Supreme Soviet speech suggests that the expansion began somewhat later. He stated that the May 1945 high of 11,365,000 was cut to 2,874,000 by 1948 (*Pravda*, 15 January 1960).

the production of consumer goods," and he reiterated the state's "special concern" for increasing consumer goods output. He did indicate that some difficulties had arisen in this policy, noting that the 1946 drought had forced postponement of the end of rationing until 1947. Turning to foreign affairs, Zhdanov was not as optimistic as in February, for he now cited the atomic danger and mentioned threats of war, referring especially to statements made by Winston Churchill. Nevertheless, in assessing this speech it is important to note that Zhdanov's protégé M. T. Iovchuk later appeared to interpret the speech as basically optimistic in its reading of the foreign situation. In a February 1976 article (*Kommunist,* no. 3, p. 85), Iovchuk recalled that in that November 1946 speech Zhdanov had voiced assurance that "no matter what forces oppose the establishment of firm general peace and security, this will finally be successfully accomplished," because "forces working for peace" were becoming stronger and stronger. Iovchuk argues that this statement was quite remarkable, since it came "at the height of the 'cold war' unleashed by the U.S. and other imperialist powers, and under conditions of atomic blackmail and threats against the USSR and other countries of East Europe." Iovchuk concludes that looking back on documents and speeches of this period, one can see that the policy of détente was not something first developed only in the 1970s.

✷ Stalin himself was still apparently optimistic about the international situation, despite Churchill's March 1946 speech announcing the establishment of an "Iron Curtain" and various other signs of international tension. Alexander Werth, in his detailed account of postwar events, cites Stalin's optimistic statements in late 1946[4] and argues that it was only in mid-1947—after announcement of the Truman Doctrine and the Marshall Plan and the breakdown of negotiations over a peace treaty for Germany and Austria—that Stalin unmistakably decided that a cold war with the West was

[4] Stalin, in answer to Werth's September 1946 questions, stated that "I do not believe in the danger of a 'new war,'" and "I do not doubt that the possibilities of peaceful cooperation, far from decreasing, may even grow" (Werth, *Russia: The Post-War Years,* pp. 142, 144). When another reporter asked Stalin a month later whether he agreed with the statement by U.S. Secretary of State Byrnes that tensions between the United States and Soviet Union were growing, Stalin said no (ibid., p. 145). And in a 21 December 1946 talk with Elliot Roosevelt, Stalin stressed his belief in peaceful coexistence with the United States, emphasized the need for more exchanges and cooperation, and argued that people were simply tired of war and that no state could start a new war now (ibid., pp. 146–47). Stalin told Roosevelt that just

unavoidable. In his 9 April 1947 talk with Harold Stassen, Stalin still appeared quite confident of peaceful coexistence and even economic collaboration with the United States.[5] As late as May 1947, the Soviet government appeared to show great confidence in the durability of peace: the 26 May decree ending the death penalty declared that the international situation after the war "had shown that the cause of peace can be considered assured for a lengthy time, despite the attempts of aggressive elements to provoke war," and therefore the death penalty was no longer needed (*Pravda*, 27 May 1947). Only later in 1947 did Stalin decide to create the Cominform, at the founding of which in September 1947 Zhdanov announced the division of the world into two camps.

Even though he announced the two camps thesis, Zhdanov may have suffered for his earlier public enthusiasm for détente and consumerism during this period when Stalin swung to hostility toward the West and placed renewed stress on defense. Zhdanov would have been especially vulnerable under such circumstances, for he had special responsibility for foreign affairs. His September 1947 Cominform speech may have been more a reflection of the Soviet leadership's position than of Zhdanov's own personal preferences; as one official later said, the speech "laid out the scientific analysis of the postwar world situation collectively worked out by the CC VKP(b) and its Politburo."[6]

Zhdanov's Clique in Ideology

Another key to understanding Zhdanov's position is obtained by examining the clique put in control of ideological organs after Zhdanov resumed his prewar position of CC secretary for ideology in 1944. Zhdanov displaced the dominant clique of ideologists led by the ultraconservatives P. F. Yudin and M. B. Mitin and advanced a younger, more flexible group led by P. N. Fedoseyev and M. T.

---

some "misunderstandings" had arisen between the Soviet Union and the West but that he saw nothing "frightful in the sense of breaches of peace or military conflict."

[5] The Soviet text of the Stassen interview, printed in the 8 May 1947 *Pravda*, as well as the texts of Stalin's answers to Werth in September 1946 and to Bailey in October 1946 and his December talk with Elliot Roosevelt, are reprinted in I. V. Stalin, *Sochineniya*, vol. 3 (XVI), 1946–53, ed. Robert H. McNeal (Stanford: The Hoover Institution on War, Revolution and Peace, Stanford University, 1967).

[6] P. A. Rodionov, first deputy director of the Institute of Marxism-Leninism, in a 10 March 1976 *Pravda* article on the eightieth anniversary of Zhdanov's birth. Rodionov also pointed out Zhdanov's special responsibility for foreign affairs.

Iovchuk. The more moderate orientation of the Zhdanov-backed ideologists emerged clearly during the 1947 philosophy debate, which will be dealt with in detail in Chapter 2.

A bitter feud developed between the two cliques and this feud played an important role in the downfall of Zhdanov himself. Yudin and Mitin had enjoyed a virtual monopoly of posts, running the Institute of Marx-Engels-Lenin, the Institute of Philosophy, the journal *Pod znamenem marksizma,* and to a considerable extent, the CC's main ideological organ, *Bolshevik.* Together with Agitprop chief G. F. Aleksandrov, a long-time Zhdanov protégé, the new people who replaced them soon established their own monopoly over ideological organs and acted as Zhdanov's principal agents in his 1946–47 conflict with Malenkov. Yudin and Mitin, embittered at Zhdanov, apparently looked to Zhdanov's enemies, probably Malenkov and Beriya, and played key roles in the two disputes that brought Zhdanov's final decline. Yudin, the Soviet representative in Belgrade, aggravated the split with Tito which seriously damaged Zhdanov, and Mitin intrigued with Lysenko to discredit Zhdanov in Stalin's eyes. Fedoseyev, Iovchuk, and the other Zhdanov protégés were ousted by the conservatives after Zhdanov's death in 1948.

The purge of the Yudin-Mitin group had begun in mid-1944, when the CC condemned a history of philosophy edited by them and also attacked the work of their journal, *Pod znamenem marksizma.* Fedoseyev was assigned to help write a revised version of the philosophy book and Iovchuk was named to replace Mitin as the journal's chief editor. Yudin was removed as director of the Institute of Philosophy and Mitin as deputy director of the Institute of Philosophy and as director of the Institute of Marx-Engels-Lenin. In 1945, when the work of the journal *Bolshevik* was also censured by the CC, Fedoseyev was named responsible secretary and later chief editor of this journal and Yudin and Mitin were dropped from the editorial board. Yudin was also attacked by Zhdanov's mouthpiece *Kultura i zhizn* in July 1946, and was fired as head of the Association of State Publishing Houses in October 1946.[7]

Though it cannot be proven that Zhdanov initiated the mid-1944 attacks on Yudin and Mitin, two circumstances suggest his influence: Zhdanov left Leningrad and moved to Moscow to take up

---

[7] The 1944–46 moves against Yudin and Mitin are explained in more detail in Chapter 2.

duties as CC secretary in charge of ideology around the middle of 1944, and Fedoseyev and Iovchuk, later purged as Zhdanov's protégés, benefited directly from the attacks on Yudin and Mitin. The exact date of Zhdanov's return to Moscow is difficult to establish, since the Leningrad papers for 1944 are apparently unavailable in the West. *Pravda's* last reference to Zhdanov in Leningrad appears to have been on 26 April 1944, and Zhdanov's biographies in some encyclopedias (*Sovetskaya voyennaya entsiklopediya* and the third edition of the *Bolshaya sovetskaya entsiklopediya*) state that Zhdanov was a member of the military council of the Leningrad front until August 1944. His biographies simply indicate that sometime in 1944 he was transferred to Moscow to work as CC secretary in charge of ideological questions.[8]

Whatever Zhdanov's role in mid-1944, he was sidetracked again in September when Finland agreed to a cease-fire. Zhdanov, who had directed the Leningrad area military front and recently been named colonel general (*Pravda*, 19 June 1944), was given the honor of signing the agreement on behalf of the USSR and named chairman of the Allied Control Commission for Finland. In this capacity, he soon moved to Helsinki to supervise Finland's compliance with the terms of the armistice agreement.[9] He was still there in December 1944 and January 1945.[10]

[8] Harrison Salisbury, in *The 900 Days* ([New York, 1969], p. 576), also states that the exact date of his transfer to Moscow cannot be fixed but was sometime after April. Statements in Zhdanov's biographies in several Soviet encyclopedias indicate that Zhdanov became CC secretary at this time, implying that he had lost this post during the war. (He had been elected CC secretary at the 18th Party Congress in 1939.) The *Sovetskaya voyennaya entsiklopediya* states that "starting 1944, he worked as secretary of the CC VKP(b) for ideological questions," and the *Ukrainska radyanska entsiklopediya* says that "starting 1944, Zhdanov worked as secretary of the CC VKP(b)." The third edition of the *Bolshaya sovetskaya entsiklopediya* states that "starting 1944, he worked in Moscow as secretary of the CC VKP(b) handling ideological questions," and the second edition of the *Bolshaya* and the *Sovetskaya istoricheskaya entsiklopediya* state that "starting 1944, he worked in the CC VKP(b)." On the other hand, M. T. Iovchuk, writing in a February 1976 *Kommunist* (no. 3, p. 82), suggests that Zhdanov might simply have resumed his duties: "Starting 1944, A. Zhdanov concentrated on work in the Central Committee of the party." Some other sources refer to Zhdanov as CC secretary during the war. A. F. Kozmin and N. D. Khudyakova, writing in the January 1964 *Voprosy istorii KPSS* (p. 27), refer to Zhdanov as a CC secretary in 1943, and Yu. P. Petrov, in his history of party-army relations (*Stroitelstvo politorganov, partiynykh i komsomolskikh organizatsiy armii i flota* [The development of political organs, party and Komsomol organizations of the army and navy], 1968, p. 301), refers to him as CC secretary in early 1942.

[9] His arrival in Helsinki was reported in the 6 October 1944 *Pravda*.

[10] He was reported in Helsinki in the 15 December 1944 *Pravda*. He is cited as

Zhdanov's period of service in Finland may have marked a political setback for him, for he appears to have played little role in leadership politics for several months. Not only did he fail to participate in Moscow political functions,[11] but he even failed to appear at the 27 January 1945 ceremony honoring Leningrad's heroism during the war—this despite the fact that he was located in nearby Helsinki. Zhdanov was formally relieved of his titles of first secretary of Leningrad city and *oblast* only days before the ceremony, perhaps to prevent him from playing the central role in the celebrations. On 26 January 1945, *Pravda* reported that a recent joint plenum of the Leningrad *oblast* and city party committees, held "in connection with the CC VKP(b) decision on releasing comrade A. A. Zhdanov from the duties of first secretary of the Leningrad *oblast* and city committees of the VKP(b) in view of his preoccupation with work in the CC VKP(b) in Moscow and in the Allied Control Commission in Finland," had elected Second Secretary A. A. Kuznetsov to succeed him.[12] As indicated above, Zhdanov had been out of Leningrad for several months and Kuznetsov had been acting as first secretary, but the change occurred only on the eve of the ceremony honoring Leningrad's successful wartime resistance. Moreover, the main *Pravda* article on the occasion appeared to shortchange Zhdanov. In this 27 January article, Zhdanov's own top deputy and successor as Leningrad first secretary, A. A. Kuznetsov, bowed to Zhdanov as head of the Leningrad party that led the local war effort but also attributed the city's successful defense partly to Molotov and Malenkov, who, the author noted, had been sent there in September 1941 to organize the defense in "the most sharp and tense period of struggle."[13]

Zhdanov appears to have left Helsinki and permanently resumed his CC duties in Moscow sometime in early 1945, and he soon moved to tighten his control over the CC's main ideological organ,

talking with Finnish leader Mannerheim during January 1945 in Max Jakobson's *Finnish Neutrality* (London, 1968), p. 37.

[11] He did not attend the 21 January 1945 Lenin anniversary ceremony in Moscow with the other Politburo members.

[12] In an article in *Voprosy istorii KPSS* (no. 1, 1961, p. 73), S. P. Knyazev, director of the Leningrad party history institute, states that Kuznetsov was elected at a 15–17 January 1945 joint plenum, and F. Leonov, in a 20 February 1965 *Pravda* article on Kuznetsov, also mentions a 17 January 1945 joint plenum.

[13] Zhdanov fared better in N. Tikhonov's 28 January 1945 *Pravda* article on Leningrad's victory, being described there as the "glorious pupil of the great Stalin." There was no reference to Molotov or Malenkov.

Bolshevik.[14] After a mid-1945 CC decree criticized the work of *Bolshevik,*[15] the journal's editorial board was purged of those apparently hostile to Zhdanov's faction. Sometime between August and October 1945 Mitin and Yudin were dropped from the board and eight new members were added, including three deputy chiefs of Agitprop (Fedoseyev, Iovchuk, and I. I. Kuzminov).[16] Fedoseyev became head of the editorial board,[17] and Kuzminov became deputy chief editor.[18] Only Aleksandrov, P. N. Pospelov, and L. F. Ilichev were retained from the old board and they appeared to have links to Zhdanov (see below). Reinforcing the notion that this was a move by Zhdanov's group is the fact that almost all the new additions were purged from the board shortly after Zhdanov's death.[19]

By mid-1946 Zhdanov had firm control of the CC's Propaganda and Agitation Administration (Agitprop) and key press organs, through *Bolshevik* chief editor Fedoseyev, *Pravda* chief editor Pospelov, Agitprop chief Aleksandrov and deputy chiefs Fedoseyev and Iovchuk, and other Agitprop officials such as S. M. Kovalev, head of the Propaganda Section.[20] These men led Zhdanov's 1946 ideological campaign and later, as Zhdanov fell into disfavor, were demoted because of their ties to him.

[14] He attended the 24 April 1945 session of the USSR Supreme Soviet, the 5 June 1945 session of the RSFSR Supreme Soviet, and the 1 May and 24 June 1945 parades.

[15] Mentioned in the editorial in the September 1945 *Bolshevik* (nos. 17–18, p. 9). The CC ordered *Bolshevik*'s editors to correct their errors and turn *Bolshevik* "in fact" into the party's theoretical organ.

[16] Issue no. 15 (August) of *Bolshevik* was signed to press on 20 August with G. F. Aleksandrov, N. A. Voznesenskiy, L. F. Ilichev, M. B. Mitin, P. N. Pospelov, and P. F. Yudin listed on the board. The next issue—no. 16, also for August—was finally signed to press only on 9 October and listed a new editorial board consisting of Fedoseyev, G. M. Gak, Iovchuk, M. Korneyev, V. S. Kruzhkov, the director of the Institute of Marx-Engels-Lenin, Kuzminov, V. P. Potemkin, the RSFSR people's commissar of education, V. I. Svetlov, the director of the Institute of Philosophy, and holdovers Aleksandrov, Ilichev, and Pospelov.

[17] He was listed as "responsible secretary" of the board in the August issue no. 16. He was listed as "chief editor" in the December issue (nos. 23–24, signed to press on 5 January 1946).

[18] Kuzminov was not so identified in the journal but his obituary in the November 1979 *Voprosy ekonomiki* suggests that he assumed this post when added to the board in 1945. It lists him as consultant, then sector head, then deputy chief of Agitprop, then deputy chief editor of *Bolshevik, 1939–49.*

[19] The June 1949 issue of *Bolshevik* dropped Fedoseyev, Aleksandrov, Gak, Iovchuk, Korneyev, Kuzminov, Svetlov, and N. Fominov from the board. Of those added in 1945, only Kruzhkov was retained in 1949. (Potemkin had died in early 1946—see *Bolshevik,* no. 4, February 1946.)

[20] For a chart of Agitprop leaders, see Appendix 3b. The journal *Pod znamenem marksizma* had been abolished in 1944.

Of these men, the closest to Zhdanov was apparently Fedoseyev, who may have developed these ties as early as 1924-34, the period during which Zhdanov was first secretary for Gorkiy *oblast*. Fedoseyev was born in Gorkiy in 1908 and graduated from Gorkiy Pedagogical Institute in 1930, initially becoming a teacher.[21] After several years of work in the Institute of Philosophy, Fedoseyev was brought into the CC apparat in 1941. At the end of 1945, when he was only thirty-seven, Fedoseyev was made chief editor of *Bolshevik*, and, by early 1946, he was also deputy chief of Agitprop.[22] Not only was Fedoseyev apparently used against Yudin and Mitin, but he played a key role in Zhdanov's campaign against Malenkov, exposing the ideological shortcomings in the Ukraine which constituted evidence of Malenkov's laxity in administering the party apparat. Yudin attacked Fedoseyev during the 1947 philosophy debate, accusing him of monopolizing ideological leadership posts, and he was forced to relinquish some of his titles (see Chapter 2). During the 1949 purge of Zhdanov protégés, Fedoseyev was ousted as chief editor of *Bolshevik* and condemned by a CC decree "On the Journal *Bolshevik*" for printing articles praising the works of Zhdanov's ally N. A. Voznesenskiy.[23] From 1950 to 1954, he held only the title of CC inspector,[24] but he staged a comeback in 1954–55 as Malenkov's influence waned and Khrushchev began promoting Zhdanovites.[25] In 1954–55 articles, he attacked Malenkov's economic policies and also some of his own conservative foes of 1949.[26]

[21] For biographical details, see the entries on Fedoseyev in the *Bolshaya sovetskaya entsiklopediya*, 3d ed., and *Deputaty verkhovnogo soveta SSSR* [Deputies of the USSR Supreme Soviet], 1970 and 1974.

[22] So identified in the 13 January 1946 *Pravda Ukrainy*.

[23] The 13 July 1949 CC decree was revealed in Suslov's attack on Fedoseyev in the 24 December 1952 *Pravda*.

[24] *Deputaty verkhovnogo soveta SSSR,* 1970, 1974.

[25] He became chief editor of the CC journal *Partiynaya zhizn* when it was reestablished in 1954 and director of the Institute of Philosophy in 1955. (The CC cadres journal, *Partiynaya zhizn,* had been created by Zhdanov in late 1946 as part of his anti-Malenkov moves and had been abolished by Malenkov as soon as he regained control over the apparat in 1948. When Khrushchev became CC first secretary, he quickly restored the journal, in a move which probably also had anti-Malenkov implications.)

[26] See Chapter 5. By the late 1960s and 1970s, Fedoseyev was again at the top of the ideological establishment. In 1967 he became director of the Institute of Marxism-Leninism, and in the 1970–71 reorganization of social science leadership, Fedoseyev became the top supervisor of social sciences, being elected Academy of Sciences vice-president for social sciences and chief of the academy's Social Sciences Section in 1971. In the late 1960s and 1970s, however, he has generally sided with conservatives

Agitprop deputy chief M. T. Iovchuk was another young protégé of Zhdanov, born like Fedoseyev in 1908 and also joining Agitprop in 1941. In 1944, probably while Zhdanov was supervising ideology, he was promoted to deputy chief of Agitprop.[27] One of his first jobs was to replace Mitin as chief editor of *Pod znamenem marksizma* and close down the journal.[28] During 1946 he promoted Zhdanov's ideological campaign at a number of conferences, and in early 1947, when Malenkov's protégés in the Belorussian leadership were ousted, Iovchuk was elected propaganda secretary for Belorussia. During the 1949 purge of Zhdanovites, he lost this post and was sent to perform academic work in the Academy of Social Sciences and Ural State University. In the mid-1950s, Iovchuk was able to return to Moscow, first to academic work at Moscow State University and then, starting in 1957, as head of a sector in the Institute of Philosophy.[29] He managed to return to top leadership in ideology in late 1970, when he was appointed rector of the CC's Academy of Social Sciences during a reorganization of that institution.[30]

Also connected with Zhdanov was P. N. Pospelov, who had been Zhdanov's top assistant in Agitprop before becoming *Pravda* chief editor in September 1940.[31] According to his biography in the

---

against moderates and has appeared on good terms with Suslov, despite their past conflicts.

[27] See his biographical entries in the *Bolshaya sovetskaya entsiklopediya*, 3d ed., and the *Bolshaya*'s 1971 yearbook. He was identified as deputy chief of Agitprop in a 9 August 1944 CC decree on Tatar ideological work (*KPSS v rezolyutsiyakh i resheniyakh syezdov, konferentsiy i plenumov TsK*, vol. 6, 1971, p. 113).

[28] See Chapter 2.

[29] In the mid-1960s he was identified as head of the sector for preparing a world history of philosophy (*Voprosy filosofii*, no. 2, 1965, p. 190) and later as head of the sector for problems of the Leninist stage of Marxist philosophy at the institute (*Voprosy filosofii*, no. 10, 1967, p. 190).

[30] During the 1970s, like his old colleague Fedoseyev, Iovchuk was again one of the most powerful ideologists — at least among those below the level of the CC secretaries Suslov and Demichev and CC Science Section head S. P. Trapeznikov. However, conflict appeared to develop between Fedoseyev and Iovchuk and in 1978 Iovchuk was suddenly retired. In one sign of the conflict, Fedoseyev attacked M. N. Rutkevich, the director of the Institute of Sociological Research, at an April 1976 meeting of the Presidium of the Academy of Sciences (*Vestnik akademii nauk SSSR*, November 1976, p. 18) and soon obtained his ouster. At the same meeting, Iovchuk alone had defended Rutkevich, with whom he had worked in Ural University, and after Rutkevich's ouster, Iovchuk hired him as head of the department of scientific communism at his Academy of Social Sciences. The Academy of Social Sciences was reorganized in February-March 1978, and both Iovchuk and Rutkevich lost their positions.

[31] His appointment as *Pravda* chief editor was announced in the 7 September 1940 *Pravda*.

Bolshaya sovetskaya entsiklopediya (2d edition), he had been deputy Agitprop chief from 1937 to 1940, and his biography in a 1972 journal[32] listed him as deputy head of the Propaganda and Agitation Section, then first deputy chief of Agitprop after it was changed to an administration.[33] During the purge of Zhdanovites in 1949, Pospelov was removed as *Pravda* chief editor and demoted to the directorship of the Institute of Marx-Engels-Lenin.[34] Elected CC secretary upon Stalin's death in March 1953, he allied himself with Khrushchev in the post-Stalin struggle between Khrushchev and Malenkov.[35]

Another young Zhdanov protégé who was later purged from Agitprop was S. M. Kovalev, head of the Propaganda and Agitation Administration's Propaganda Section.[36] Playing an active role in Zhdanov's 1946 campaign, he wrote several articles, including the key 20 July 1946 *Kultura i zhizn* piece denouncing Ukrainian nationalist deviations and spoke at a Ukrainian ideological conference that is reported in the 30 May 1947 *Kultura i zhizn.* Replaced during the reorganization of Agitprop in July 1948[37] he too staged a comeback many years later, becoming *Pravda*'s editor for Marxist-

[32] *Novaya i noveyshaya istoriya,* no. 4, 1972, p. 207.

[33] Agitprop was made into an administration (*upravleniye*) by the 20 March 1939 resolution of the 18th Party Congress (*KPSS v rezolyutsiyakh i resheniyakh syezdov, konferentsiy i plenumov TsK,* pt. 3, 7th ed., 1954, p. 372).

[34] His biography in the *Bolshaya sovetskaya entsiklopediya,* 2d ed.

[35] In 1954 he was selected to head the special commission that investigated Stalin's repressions, as Khrushchev revealed, and prepared the material that Khrushchev used for his secret speech exposing Stalin's crimes at the 1956 party congress—a report opposed by Malenkov, Molotov, Kaganovich, and Voroshilov (*Khrushchev Remembers* [Boston, 1970], pp. 345–50). In 1960 he was dropped as CC secretary and in 1961 he became director of the Institute of Marxism-Leninism. Despite his association with Zhdanov and Khrushchev and his involvement with rehabilitations of Stalin's victims, Pospelov has a well-established reputation as a conservative, based on his defense of Stalin and his heavy-handed attacks on "liberals" in social sciences throughout the years. Some examples of his resistance to destalinization and liberalization in historiography are cited in Aleksandr Nekrich's *Otreshis ot strakha* [Renounce fear], (London, 1979), pages 64, 176, 179, 285, 286, 330, 340, 341, and in the excerpt from that book which appeared in the fall 1977–78 issue of *Survey* (no. 105), pages 140–41.

[36] Born in 1913, according to the December 1971 *Zhurnalist,* he was only thirty-three in 1946.

[37] He was last identified as head of the Propaganda Section in the 21 April 1948 issue of *Kultura i zhizn.* In mid-1948 the Propaganda and Agitation Administration became the Propaganda and Agitation Section and the administration's sections (*otdely*) became sectors (*sektory*). P. N. Lyashchenko was identified as head of the sector for party propaganda in the 31 July 1949 *Kultura i zhizn,* apparently succeeding Kovalev, who was transferred to a publishing house. The emigré historian

Leninist theory in the 1960s and attaining notoriety in 1968 by writing the long *Pravda* article (26 September) that set forth the so-called Brezhnev Doctrine of limited sovereignty—that is, justified the Soviet invasion of Czechoslovakia.

The position of G. F. Aleksandrov, the top figure in ideology under Zhdanov, is more complicated than that of Fedoseyev and the others. Brought into Agitprop and promoted to its chief post by Zhdanov, he appeared to be a key member of the Zhdanov group during the 1946–47 conflict, but later switched sides and became a Malenkov protégé. Aleksandrov had become an assistant to Agitprop chief Zhdanov by the age of thirty and had succeeded Zhdanov as chief of the administration only two years later.[38] He remained in this post throughout the war and in 1946 was promoted to the leadership's inner circle, being elected an Orgburo member at the March 1946 plenum.

Some observers have felt that he defected to Malenkov's side during this period, when Zhdanov was isolated in Leningrad and Malenkov dominated the apparat.[39] However, he still seems to have been on Zhdanov's side in 1946 and was protected by Zhdanov's clique through early 1947 and, at least in part, during the June 1947 debate as well (see Chapter 2). As Agitprop chief, he supervised the Agitprop newspaper *Kultura i zhizn* which was set up in mid-1946 and which became the main instrument for Zhdanov's crackdown. Eventually he was specifically identified as its chief editor.[40]

Zhdanov criticized Aleksandrov's book on the history of West European philosophy in June 1947, however, and Aleksandrov was afterwards demoted to director of the Institute of Philosophy. He appears to have changed sides at that point. He joined in the attacks

---

Aleksandr Nekrich mentions Kovalev as acting director of *Gospolitizdat* (the State Publishing House for Political Literature) in 1948 or 1949 (Nekrich, *Otreshis ot strakha*, p. 97). In 1951, Kovalev became director of *Politizdat* (the Publishing House for Political Literature), according to his biography in the December 1971 *Zhurnalist*.

[38] A sketchy biography of Aleksandrov can be found in the *Filosofskaya entsiklopediya*, vol. 1, 1960. *Pravda* on 7 September 1940 announced Aleksandrov's appointment as chief of Agitprop, succeeding Zhdanov, who, it announced, would continue to have "supervision" over it.

[39] In particular, see Boris Nikolayevskiy's "Na komandnykh vysotakh Kremlya." [On the commanding heights of the Kremlin], *Sotsialisticheskiy vestnik*, 21 June 1946.

[40] An anonymous "editorial collegium" signed the newspaper until 20 July 1947, when Aleksandrov began signing as chief editor. It reverted to "editorial collegium" again on 21 September 1947.

on his former Zhdanovite colleagues during 1948–49, and after Malenkov became premier in 1953, he appointed Aleksandrov his culture minister in early 1954. When Malenkov was removed as premier in early 1955, Aleksandrov was also forced out, and in a September 1955 *Kommunist* article his former ally B. M. Kedrov attacked him for the same mistakes that Zhdanov had pointed out in 1947.

Kedrov, who became deputy director of the Institute of Philosophy in 1945 and who was selected to become chief editor of the new journal *Voprosy filosofii* in 1947, must also be considered a member of the Zhdanov clique. He was the most outspoken liberal of the group, and by late 1947 had become the main target of the dogmatists. Whereas many of the other Zhdanov ideologists later wound up on the conservative side in the clashes of the 1960s and 1970s (especially Fedoseyev and Pospelov), Kedrov remained consistently on the side of reform. His activities, public stands, and fate (described in Chapters 2 and 6) provide the most revealing evidence of the struggle over moderation during the postwar Zhdanov era.

Struggle with Malenkov

The sharp personal rivalry between Zhdanov and Malenkov is one of the best understood features of postwar Soviet politics. Malenkov, the specialist in matters of cadre selection and the work of party organs, had become quite powerful during the war, administering the party apparat in Stalin's stead. Zhdanov, the supervisor of ideological work before the war, was then located in besieged Leningrad, and both because of his physical absence from the center and because the initiation of important personnel appointments was in Malenkov's hands, he appeared to fall behind Malenkov in influence in the central and regional party apparats. In addition to their personal competition for power, Zhdanov and Malenkov had certain policy differences. As indicated above, Zhdanov seemed to take a more moderate position on questions of defense and economic priorities than Malenkov, at least in 1946. Moreover, they had long differed over how to organize the CC apparat and over whether ideology or practical economic questions should dominate party work. Malenkov advocated the organization of CC departments by industrial branches in a system somewhat parallel to the ministerial structure—a system that would permit CC officials to supervise ministries and economic organs more closely, in fact allowing party organs to concentrate on economic matters. Zhdanov urged a structure based mainly on two big departments, one for ideology and one for cadres, which would focus CC work on these subjects and leave economic administration more to the ministries and economic administrators. Zhdanov's approach would give ideology equal weight in party work, whereas Malenkov's downplayed ideology (and hence the usefulness of Zhdanov and his apparat).

The Zhdanov-Malenkov struggle was intensified by the reorganization of Soviet leadership organs in the wake of the war. The State Defense Committee (*Gosudarstvennyy komitet oborony*), which in effect ran the country during World War II, was abolished in late 1945, and the Politburo and other top party organs resumed their normal functions. However, the individuals dominating the State Defense Committee were not the same as those most prominent in the Politburo. Junior leaders Malenkov and Beriya, who had become the two key members of the committee other than Stalin, were only candidate members of the Politburo, while some senior Politburo members, such as Zhdanov and Andreyev, did not sit on the committee. In March 1946, all the top organs—the Politburo, Orgburo, and Secretariat in the party, and the Council of Ministers in the government—were reorganized. Malenkov and Beriya became full members of the Politburo and, even though two Zhdanov protégés (or allies), A. A. Kuznetsov and G. M. Popov, were added to the Secretariat, the reshuffle made Malenkov the equal to Zhdanov as Stalin's top party deputy. In fact, Malenkov was even listed ahead of Zhdanov in the reorganization, implying a decline in Zhdanov's influence. Given this arrangement, the intensified rivalry between Stalin's two top deputies in the CC Secretariat soon dominated power relationships.

The State Defense Committee was abolished on 4 September 1945, marking the return to normal peacetime leadership by party and government organs. When the committee was established back in June 1941, it had been given "full power in the state" in order to direct the war effort.[41] Most official sources define it as a state organ, but a few claim that it exercised party leadership functions as well. Most studies, especially the more recent official histories, assert that

---

[41] *Bolshaya sovetskaya entsiklopediya* (2d ed., vol. 12, p. 317), quoting Stalin's 3 July 1941 speech. Other encyclopedias and CPSU histories also use the "full power" phrase.

the committee acted under the direction of the Politburo, but such assertions may represent an attempt to upgrade the Politburo's role after the fact and to maintain the fiction that the regularly constituted organs of party leadership continued to play their proper role even during the war.[42] At the very least, it is clear that Soviet historians confront difficulties in any attempt to define the relationship between the State Defense Committee and the Politburo.[43]

Whether or not it had formal jurisdiction over party organs, the State Defense Committee appears to have acted as the real leadership body during the war and its abolition coincided with the restoration of Politburo functions.[44] The committee was headed by Stalin, with

[42] Yu. P. Petrov, perhaps the leading historian on party-military relations during the war and a coeditor of the official party history of this period (*Istoriya kommunisticheskoy partii sovetskogo soyuza* [History of the Communist Party of the Soviet Union], vol. 5, bk. 1, 1938–1945, published in 1970), has gone farthest in asserting a party role for the State Defense Committee. In his *Stroitelstvo politorganov, partiynykh i komsomolskikh organizatsiy armii i flota* (p. 272), Petrov declared that the State Defense Committee "united in itself party leadership and soviet executive power" and that party, soviet, military, trade union, and Komsomol organizations were all obliged to carry out its orders. He had made basically the same statement on page 347 of his 1964 book, *Partiynoye stroitelstvo v sovetskoy armii i flote* [Party development in the Soviet Army and Navy]. In an article in the May 1970 *Voprosy istorii,* Petrov wrote that the creation of the State Defense Committee "could not help but introduce changes in the existing peacetime system of activities of the appropriate soviet and also party organs," and that the State Defense Committee "was allotted certain party rights" and often had party secretaries as its plenipotentiaries (p. 15). At the same time, he complained that some historical works downplayed the fact that the committee "worked under the leadership and control of the party," and he declared that the Politburo, Orgburo, and Secretariat had held two hundred meetings during the war and "decided all very important questions of leading the country and conducting the war." This latter statement—about the supreme party organs—was included in the party's official history for this period (*Istoriya kommunisticheskoy partii sovetskogo soyuza,* vol. 5, bk. 1, covering 1938–45, published in 1970, p. 642) and has been cited by various historical articles, such as that by A. M. Iovlev in the April 1975 *Voprosy istorii KPSS* (p. 58) and V. V. Kazarinov in the May 1975 *Voprosy istorii KPSS* (p. 123). Petrov and other writers note joint meetings of the Politburo and State Defense Committee. Another frequent writer on this subject, Gen. Ye. Ye. Maltsev, wrote in the 4 April 1975 *Pravda* that the committee "united the functions of state and party leadership and full power in the country." He also added, however, that "all principled questions of conduct of the war were decided by the party CC—the Politburo, Orgburo, and Secretariat," and in a May 1975 *Voprosy istorii KPSS* (p. 9) article coauthored with Ye. F. Nikitin, he stated that the State Defense Committee's activities "were directed by the Politburo."

[43] D. M. Kukin, deputy director of the Institute of Marxism-Leninism, complained in a June 1971 *Voprosy istorii KPSS* article (p. 55) that the wartime roles of the Politburo, Orgburo, Secretariat, and State Defense Committee needed much more examination by historians.

[44] According to Yu. P. Petrov, who apparently had access to archives of Soviet leadership organs, the abolition of the committee coincided with restoration of the

Molotov as deputy chairman, and Malenkov, Beriya, Voroshilov, and later Voznesenskiy, Kaganovich, Mikoyan, and Bulganin as members.[45] Several of the nine full members of the Politburo (Zhdanov, Andreyev, Khrushchev, and Kalinin) were excluded and the committee's key figures (Malenkov, Beriya, and Voznesenskiy) were only candidate members of the Politburo. The Politburo's ability to function as the supreme leadership body during the war was complicated by the fact that some of its members were stationed outside Moscow and were probably not available for its frequent meetings: Zhdanov was in Leningrad, Khrushchev in the Ukraine, and Andreyev in Kuybyshev.[46]

The power and influence of Beriya and Malenkov were all the greater since some of the other members of the State Defense Committee and Politburo suffered serious political setbacks during the war. Kaganovich was openly censured for failure and was fired as people's commissar for railroads in 1942.[47] During the Nazi advance on Leningrad in 1941, Stalin angrily accused Voroshilov and Zhdanov, who were then in charge of Leningrad defenses, of being "specialists in retreat."[48] Voroshilov soon was stripped of command

functions of the Politburo and other party leadership bodies (Petrov, *Stroitelstvo*, 1968, p. 389). Petrov also states that with the end of the war "it was necessary to abolish the extraordinary forms of leadership of the country, end the militarization of party life and certain restrictions on intraparty democracy caused by the war," and return to the "party leadership system set by the party statute" (ibid., p. 388). The new CPSU history issued in 1980 specified that the committee was abolished on 4 September 1945 and that a 29 December 1945 Politburo decision restored the regular holding of Politburo sessions (*Istoriya kommunisticheskoy partii sovetskogo soyuza*, vol. 5, bk. 2, pp. 11–12).

[45] See the entries on the State Defense Committee in the *Sovetskaya voyennaya entsiklopediya* and the *Bolshaya sovetskaya entsiklopediya*, 2d and 3d editions, and also entries on these individuals in the *Bolshaya*. Bulganin was added only in 1944, according to his biography in the second edition of the *Bolshaya*.

[46] Andreyev was assigned authority over the Western Siberia–Central Asia area and was stationed in Kuybyshev (see A. F. Vasilyev in *Voprosy istorii KPSS*, August 1964, p. 51, or the *Istoriya velikoy otechestvennoy voyny sovetskogo soyuza, 1941–1945* [History of the Great Fatherland War of the Soviet Union, 1941–1945], vol. 2, p. 149.)

[47] A State Defense Committee decree (25 March 1942) declared that Kaganovich "was not able to cope with work under wartime conditions," and a Supreme Soviet Presidium ukase fired him as people's commissar, according to the *Istoriya velikoy otechestvennoy voyny sovetskogo soyuza, 1941–1945*, vol. 2, p. 528.

[48] Harrison Salisbury, in *The 900 Days* (pp. 217–18), explained that a 20 August 1941 order by Zhdanov and Voroshilov to create a Council for the Defense of Leningrad in order to organize for street-by-street resistance angered Stalin, who phoned them on 21 August and, apparently thinking they planned to abandon Leningrad, accused them of being "specialists in retreat." According to Salisbury, Stalin said that Zhdanov "concerns himself with only one thing—how to retreat."

and transferred back to Moscow in disgrace,[49] and Zhdanov also apparently suffered a serious loss of Stalin's confidence.[50]

The process of returning to peacetime leadership procedures continued through late 1945 and into 1946. On 5 October 1945 the long delayed election of a new Supreme Soviet was scheduled for February 1946, and in December 1945 a Politburo session adopted a decision to resume regular Politburo operations, including biweekly meetings.[51] During 1946, according to historian Petrov, Politburo decisions redefined the roles and procedures of the Orgburo and Secretariat, the other top organs of the CC.[52]

Another aspect of this return to regular party procedures was the preparation for a new party congress—long overdue, since the previous congress had been in 1939. According to a *Pravda* editorial of 28 April 1964 (the purpose of which was to attack the Chinese Communist Party's violations of party procedure), the postwar Politburo twice decided to call a congress and even set dates for it, "but Stalin then managed to get it postponed, arguing that he was

Aleksandr Chakovskiy, in his historical novel *Blokada* [The blockade], serialized in *Znamya*, also mentions that Stalin called Zhdanov and Voroshilov "specialists in retreat" (*Znamya*, June 1971, p. 50). The story is told in some detail in A. V. Karasev's *Leningradtsy v gody blokady, 1941–1943* [Leningraders during the blockade, 1941–1943], pp. 105–6, though without repeating Stalin's sharp accusation. Relying mainly on the version given in D. V. Pavlov's 1958 book *Leningrad v blokade* [Leningrad in the blockade], Karasev related how Stalin phoned Voroshilov and Zhdanov on 21 August expressing his "dissatisfaction" with their creation of a Council for the Defense of Leningrad "without his permission" and rejecting their explanations.

[49] Chakovskiy, in *Blokada* (*Znamya*, March 1970, pp. 135–36, and June 1971, pp. 39–41), describes Zhukov taking over from the disgraced Voroshilov and overriding a decision by Zhdanov, as well as Voroshilov, on preparations for the defense of Leningrad.

[50] Chakovskiy writes that after Zhdanov and Voroshilov had received repeated reproaches for poor leadership of military operations and had been called "specialists in retreat," the "personal relations between Zhdanov and Stalin lost their earlier warmth" (*Znamya*, March 1973, p. 15). Moreover, shortly after Stalin's late August rebuke to Zhdanov, Malenkov and Molotov were sent to Leningrad to organize its defenses. (A 27 January 1945 *Pravda* article mentioned this Malenkov-Molotov mission as occurring in early September, whereas the official party history [*Istoriya kommunisticheskoy partii sovetskogo soyuza*, vol. 5, bk. 1, p. 220] says that a commission including Molotov, Malenkov, Kosygin, and others was sent to Leningrad at the end of August).

[51] Petrov, *Stroitelstvo*, 1968, p. 389.

[52] These April and August 1946 decisions indicated that the Orgburo would direct "party and party-organizational work" and meet at least once a week, whereas the Secretariat, which would meet as often as necessary, would prepare questions for Orgburo consideration and check on fulfillment of Politburo and Orgburo decisions (Petrov, ibid., p. 389).

not ready to make the CC report [to the congress] and that in view of his age it would be difficult and he needed more time to prepare."[53] The article indicated that these Politburo initiatives occurred in the 1946–48 period. Essentially the same story was related by G. I. Shitarev in the July 1964 *Voprosy istorii KPSS*. Since both articles maintain that the 1946–50 five-year plan should have been discussed and approved at a congress, and since the five-year plan was approved at a March 1946 Supreme Soviet session (the first session of the newly elected Supreme Soviet), we can assume that the congress would have been held in early 1946. Such a congress would also have been appropriate to authorize the March 1946 reorganization of party leadership, but, as it happened, that reorganization was approved only by a plenum.[54]

The abolition of the wartime power structure was capped by a general reorganization of leadership posts in March 1946, a reshuffle that considerably altered the relative power of Stalin's key deputies. In late 1945 and early 1946, prior to the reorganization, Stalin's top deputies in the party had been Zhdanov and Andreyev. Both were CC secretaries and full Politburo members. Zhdanov supervised ideology, while Andreyev, who also was minister of farming (and chairman of the Party Control Commission), ran agriculture. The remaining member of the Secretariat, Malenkov, supervised party organizational work and had great power, but was only a Politburo candidate member.[55] In the Politburo, the ranking figures, in addition to secretaries Stalin, Zhdanov, and Andreyev, were deputy

[53] Stalin appeared to be citing his advanced age already in mid-1945. According to an account by Admiral N. G. Kuznetsov, Stalin, speaking at a banquet after the 24 June 1945 victory parade, startled his military commanders by reminding them that he was going on sixty-seven and musing that "well, I will work another two-three years and then I should go" (*Neva*, May 1965, p. 161). He left for Sochi on vacation on 9 October 1945 (*Pravda*, 10 October 1945) and appeared to remain there during the remainder of 1945 and also spent much time on vacation in 1946 and 1947 (see H. Montgomery Hyde, *Stalin* [New York, 1971], pp. 550, 552, 559).

[54] The second date for a congress may have been in 1948, since Malenkov in his September 1947 Cominform speech apparently announced that a congress was being prepared. According to V. Dedijer, Malenkov spoke "in detail about preparations for a VKP(b) congress," which was to adopt a "15-year long-term plan for transition from socialism to communism" (Dedijer, *Josip Broz Tito: prilozi za biografiju* [Belgrade, 1953], p. 475).

[55] Elected in February 1941. The power accumulated by Malenkov and Beriya during the war was impressive enough to persuade one close U.S. embassy observer, political counselor George Kennan, that they, rather than Stalin, often determined Soviet actions. See the excerpts from an early 1945 letter by Kennan published in the September 1968 *Slavic Review*, pp. 481–84.

premiers Molotov, Mikoyan, and Kaganovich. The membership of the Orgburo, the CC's other leading body, is unclear, but included and was clearly dominated by the four CC secretaries, Stalin, Zhdanov, Andreyev, and Malenkov.[56] On the government side, the leaders were Premier Stalin and deputy premiers Molotov, Mikoyan, and Kaganovich. Police chief Beriya and Gosplan head Voznesenskiy were also deputy premiers but only candidate members of the Politburo. Beriya had given up direct leadership of the police in January 1946 when S. N. Kruglov became NKVD chief (*Moskovskiy bolshevik*, 15 January 1946).

In mid-March 1946 this lineup was considerably altered, most notably with the promotion of Malenkov and Beriya and the demotion of Andreyev. On 20 March *Pravda* announced that a new Council of Ministers had been named at a 19 March session of the Supreme Soviet and that a CC plenum had "recently" been held to reorganize party leadership bodies.[57] The plenum elected Malenkov and Beriya as full members of the Politburo, Kosygin and Bulganin as candidate members, and also chose a new Secretariat and Orgburo. Andreyev (an apparent ally of Zhdanov and foe of Malenkov) was dropped from the Secretariat and Orgburo,[58] while Leningrad First Secretary A. A. Kuznetsov (a close Zhdanov protégé) and Moscow First Secretary G. M. Popov (an apparent ally of Zhdanov) were added to the Secretariat and Orgburo.

Some decisions taken about the time of the March plenum were not publicized, notably the significant changes in the CC apparat. In one such change Chelyabinsk First Secretary N. S. Patolichev was named to head the Organizational-Instructor Section (*Organiza-*

---

[56] The Orgburo elected at the March 1939 congress had included nine members—Stalin, Zhdanov, Malenkov, Kaganovich, Andreyev, Mekhlis, Mikhaylov, Shvernik, and Shcherbakov—but this membership had probably changed considerably during the war and the members in February 1946 are unknown.

[57] The CPSU history issued in 1980 gives the dates of the March 1946 plenum as 11, 14 and 18 March 1946 and indicates that the Supreme Soviet session lasted from 15 to 19 March (*Istoriya kommunisticheskoy partii sovetskogo soyuza*, vol. 5, bk. 2, p. 623). This unusual pattern of overlapping dates suggests that the complicated switching of top leadership positions was unusually difficult.

[58] He was transferred to the less vital post of deputy premier—one of eight. He also lost the post of minister of farming. Khrushchev spoke of Malenkov's dislike of Andreyev in his 31 March 1961 Note to the Presidium—see N. S. Khrushchev, *Stroitelstvo kommunizma v SSSR i razvitiye selskogo khozyaystva* [Construction of communism in the USSR and the development of agriculture, hereafter referred to as Khrushchev's speeches], vol. 5, p. 349. For more on Andreyev's bad relations with Malenkov, see the last part of this chapter.

Chart A. Party leadership changes, March 1946[59]

Politburo		Secretariat	
January 1946	*March 1946*	*January 1946*	*March 1946*
Stalin	Stalin	Stalin	Stalin
Zhdanov	Zhdanov	Zhdanov	Zhdanov
Andreyev	Andreyev	Andreyev	Kuznetsov
Molotov	Molotov	Malenkov	Malenkov
Kaganovich	Kaganovich		Popov
Voroshilov	Voroshilov		
Mikoyan	Mikoyan		
Khrushchev	Khrushchev		
Kalinin	Kalinin (died		
	June 1946)		
	Malenkov		
	Beriya		

Candidate Members	
Shvernik	Shvernik
Voznesenskiy	Voznesenskiy
Malenkov	Bulganin
Beriya	Kosygin

Chart B. Government leadership changes, March 1946

	January 1946	*March 1946*
Premier	Stalin	Stalin
Deputy Premier	Molotov	Molotov
	Mikoyan	Mikoyan
	Kaganovich	Kaganovich
	Beriya	Beriya
	Voznesenskiy	Voznesenskiy
	Kosygin	Kosygin
		Voroshilov
		Andreyev
Defense Minister	Stalin	Stalin
MVD	Beriya (until	Kruglov
	14 Jan.)	
MGB	Merkulov	Merkulov
Foreign Minister	Molotov	Molotov
Foreign Trade	Mikoyan	Mikoyan
Minister		
Gosplan Chairman	Voznesenskiy	Voznesenskiy

tsionno-instruktorskiy otdel) — a decision not revealed until the
publication of Patolichev's autobiography over thirty years later.[60]
Since Malenkov's close protégé M. A. Shamberg had headed this

[59] For the Orgburo membership, see below, pp. 43–44. Its membership before
March is unclear.
[60] N. S. Patolichev, *Ispytaniye na zrelost* [Testing for maturity] (Moscow, 1977), p.
280. Patolichev also asserts that A. A. Kuznetsov had been made CC secretary
"already before the plenum" (p. 279).

section,[61] Patolichev's appointment was an important step toward bringing down Malenkov; it removed one of the two CC cadres departments (the other being the Cadres Administration) from Malenkov's control and gave it to a protégé of his foe Andreyev. Patolichev's antipathy to Malenkov is clear from his autobiography, in which Malenkov is mentioned rarely and mainly in negative contexts: as initiator of a harsh and sometimes unfair purge of *oblast* leaders in 1941 and as being removed as CC secretary in 1946 because of poor checking on local party organs.[62] In contrast, Patolichev presents Zhdanov as sympathetic[63] and repeatedly lauds the personal qualities of Andreyev, whom, as he himself said, he and other apparatchiks sought to emulate. Furthermore, he presents Andreyev as the patron who interceded for him at every key juncture in his early rise: (1) Andreyev in 1938 appointed the twenty-nine-year-old Patolichev, then an inexperienced military engineer, to the powerful post of CC instructor, thus launching his political career; (2) Andreyev at a January 1939 Orgburo meeting proposed appointing Patolichev first secretary of Yaroslavl *oblast*; (3) Andreyev in 1940 tried to talk Patolichev into accepting appointment as Komsomol CC first secretary and also helped Patolichev protect an associate from being put on trial on trumped-up charges; (4) Andreyev at the 18th VKP Conference (1941) nominated Patolichev for election as a CC member; and (5) Zhdanov and Andreyev in December 1941 informed Patolichev of his transfer from Yaroslavl leader to Chelyabinsk first secretary.[64] Although Malenkov was cadres chief throughout these years, Patolichev gives him no credit whatsoever for his rise.

During much of 1946 the Soviet press rarely identified CC department heads or spelled out the duties of Orgburo members,

[61] Shamberg was identified in this post in the 15 December 1945 *Pravda*. His son was married to Malenkov's daughter, according to Khrushchev (*Khrushchev Remembers*, pp. 292–93). On the original tapes from which the Khrushchev recollections were taken, Khrushchev stressed Malenkov's special closeness to Shamberg, stating that Shamberg had worked with Malenkov for many years in the apparat and that resolutions assigned to Malenkov for preparation "were primarily prepared by Shamberg" (Russian-language transcript of pt. I, p. 804). After the 1946 reorganization, Shamberg became a CC inspector (identified as such in a May 1947 *Partiynaya zhizn*, no. 9).

[62] Patolichev, *Ispytaniye*, pp. 113, 284.

[63] He notes, for example, that at the 1939 party congress Zhdanov pointed to Patolichev's work as a good example and used an advance copy of Patolichev's congress speech in his own speech (ibid., pp. 77, 89).

[64] See ibid., pp. 70, 81–82, 97–99, 114, 165.

perhaps because of the secret reorganization of the apparat between March and October 1946.[65] Nevertheless, part of the picture can be pieced together. The Orgburo elected at the March 1946 plenum had fifteen members, among whom the five CC secretaries—Stalin, Malenkov, Zhdanov, A. A. Kuznetsov, and Popov—clearly were senior. Also included were key CC officials—Agitprop chief Aleksandrov, Shatalin, deputy chief of the Cadres Administration, and Patolichev, head of the Organizational-Instructor Section—plus Suslov, who, according to later biographies, was transferred from head of the CC Bureau for Lithuania to unspecified work in the CC apparat in March 1946.[66]

Zhdanov appeared to have more weight than Malenkov in the Orgburo, since A. A. Kuznetsov, Rodionov, Aleksandrov, Popov, and Patolichev were allied to Zhdanov, and only Shatalin and Andrianov were close to Malenkov.[67] In sum, then, the Orgburo elected in March 1946 included the following:

I. V. Stalin, general secretary and premier

A. A. Zhdanov, CC secretary and supervisor of Agitprop

G. M. Malenkov, CC secretary and supervisor of Cadres Administration

A. A. Kuznetsov, CC secretary and, according to Khrushchev, soon also supervisor of state security[68]

G. M. Popov, CC secretary and Moscow first secretary

N. A. Bulganin, Politburo candidate member and deputy minister of defense

G. F. Aleksandrov, chief of CC Propaganda and Agitation Administration

N. N. Shatalin, deputy chief of CC Cadres Administration

[65] An example of the secrecy cloaking the apparat reorganization appeared when the Kolkhoz Affairs Council was set up. The 9 October 1946 *Pravda*, in identifying its members, listed deputy chairmen Patolichev and Andrianov only as CC secretary and Orgburo member, respectively, even though they also were—judging by what Patolichev says in his autobiography—chief and first deputy chief of the new CC Administration for Checking Party Organs.

[66] For example, his biography in the *Bolshaya sovetskaya entsiklopediya*, 3d ed.

[67] Shatalin was long Malenkov's deputy in the Cadres Administration. Andrianov was named first secretary of Leningrad after the Zhdanovites were purged in 1949, and because of his role in the Malenkov-Beriya purge of Leningrad, he was fired by Khrushchev in late 1953.

[68] N. S. Khrushchev, "The Crimes of the Stalin Era," a reprint of Khrushchev's speech to a closed session of the 20th CPSU Congress (hereafter referred to as Khrushchev's secret speech), by *The New Leader*, 16 July 1956, p. S45.

N. S. Patolichev, head of CC Organizational-Instructor Section

M. A. Suslov, unspecified CC work

N. A. Mikhaylov, Komsomol first secretary

M. I. Rodionov, Gorkiy first secretary until June 1946, then RSFSR premier

V. M. Andrianov, Sverdlovsk first secretary until sometime in mid-1946, then first deputy chief of the CC Checking Administration, according to Patolichev[69]

V. V. Kuznetsov, chairman of Trade Union Council

L. Z. Mekhlis, minister of state control

The CC cadres apparatus at this time consisted of two departments: the Cadres Administration and the Organizational-Instructor Section. Malenkov had headed the first since the late 1930s, although there are no press identifications of him in this post during 1946. His deputies were N. N. Shatalin,[70] Ye. Ye. Andreyev,[71] V. D. Nikitin,[72] and A. S. Pavlenko.[73] The Organizational-Instructor Section was headed by Malenkov's foe, Patolichev.[74]

The net result of these March 1946 changes was the creation of a precarious balance between contending forces. Malenkov and Beriya had been strengthened because of their election to the Politburo, but at the same time they had to surrender some important levers of power. Beriya relinquished direct leadership of the MVD to S. N. Kruglov, not one of his closest deputies, and Malenkov lost control of the Organizational-Instructor Section with the appointment of Andreyev's protégé Patolichev.

Malenkov's Fall

The precarious balance between Malenkov and Zhdanov arranged at the March 1946 plenum did not last long. Zhdanov soon raised questions about Malenkov's administration of the CC apparat

[69] Patolichev, *Ispytaniye*, pp. 280–84.

[70] Identified in the 13 January 1946 *Izvestiya*, and again in the 6 August 1946 *Pravda Ukrainy*.

[71] Identified as a senior CC official in the 12 November 1945 *Vedomosti verkhovnogo soveta SSSR*, and later, in *Partiynaya zhizn*, no. 2, November 1946, identified as deputy chief of this administration.

[72] Identified as such in the 12 January 1946 *Izvestiya* and in the 16 December 1946 *Pravda*. His 18 April 1959 *Pravda* obituary lists him as deputy chief of this administration starting in 1944.

[73] Long identified as sector head in this administration, and identified as deputy chief in the 2 November 1946 *Pravda*.

[74] Starting in March 1946, according to his book.

and on 4 May persuaded Stalin to suspend Malenkov as CC secretary and cadres chief. Zhdanov's men—primarily in Agitprop—then set out to expose deviations and abuses within regional party apparatuses as evidence of Malenkov's negligence. After serious evidence of such laxity was brought to light in June and July, Stalin ordered a thorough reorganization of Malenkov's cadres apparat and Malenkov's demotion was confirmed.

The first exposures came on 24 June, when Zhdanov's agent Fedoseyev went to the Ukraine to correct ideological deviations there. The second step was the publication on 28 June of the first issue of *Kultura i zhizn,* a newspaper clearly set up as an instrument for the crackdown, the inaugural issue of which included a general attack on ideological shortcomings. On 26 July, the Politburo condemned errors in cadre work by the Ukrainian leadership, threatening the position of Ukrainian leader and Politburo member N. S. Khrushchev, and also criticized party-wide laxness in admitting new members. On 2 August, the Politburo denounced mistakes by the CC apparat, especially the cadre organs, and Malenkov was transferred to deputy premier. After winning these major victories, Zhdanov, armed with a 14 August CC decree on the ideological errors of two Leningrad journals, announced the ideological campaign in speeches in his former bailiwick of Leningrad. Zhdanov thereby restored the priority of ideology which had been badly eroded during the war and established the superiority of his ideological apparat over Malenkov's cadres apparat.

Malenkov still appeared to enjoy very high standing at the May Day parade in 1946, ranking ahead of Zhdanov.[75] Yet only three days later, according to a remarkable eyewitness account given in Patolichev's autobiography, Stalin removed Malenkov from the Secretariat for errors in administering the party apparat. In his account—our only evidence that Malenkov was demoted in May rather than in August 1946—Patolichev tells of being called into a meeting with CC secretaries Stalin, Zhdanov, and Kuznetsov late on the night of 4 May. When Patolichev came in, they were discussing charges that CC control over local party organs was too loose — charges presumably raised by Zhdanov and his protégé

---

[75] *Pravda* listed the leaders as ascending to the tribune in this order: Stalin, Beriya, Malenkov, Mikoyan, Zhdanov, Andreyev, Kaganovich, Shvernik, Voznesenskiy, Bulganin, Kosygin. Molotov, normally ranked second, was absent.

Kuznetsov—and Stalin had already decided to remove Malenkov from the Secretariat. Stalin asked Patolichev, as head of the CC's Organizational-Instructor Section, whether he thought CC checking to be lax. Patolichev emphatically agreed that CC checking on local organs had become poor, whereupon Stalin suggested the creation of a new CC organ—the Administration for Checking Party Organs (*Upravleniye po proverke partiynykh organov*)—to end such negligence. Moreover, Stalin suggested making Patolichev its head and also making him CC secretary—in effect suggesting that he take over for Malenkov.[76] Patolichev's appointments as CC secretary and chief of the Organizational-Instructor Section and then of the new Checking Administration were not reported publicly at the time, and even Patolichev's later encyclopedia biographies omit most of these details.[77]

After appointing Patolichev, Stalin asked him whom he wished to have as deputies, and Patolichev listed Bashkir First Secretary S. D. Ignatyev, Sverdlovsk First Secretary V. M. Andrianov, and Kazakh First Secretary G. A. Borkov. Stalin responded by suggesting that Andrianov be first deputy and that N. M. Pegov, first secretary of Primorskiy *kray,* be a deputy chief in addition to Ignatyev and Borkov. After brief discussion, Stalin and Patolichev decided to call the other officials of the new administration "inspectors" and to select the best *oblast* secretaries for this post so that they could deal authoritatively with local organs. In the discussion, Stalin called attention to the following as best qualified: Ignatyev, Borkov, Pegov, Kuybyshev First Secretary V.G. Zhavoronkov, Perm First Secretary N. I. Gusarov, Kemerovo First Secretary S. B. Zadionchenko, Chkalov First Secretary G. A. Denisov, and V. D. Nikitin, deputy chief of the Cadres Administration.[78]

During the summer, as Patolichev began organizing the new Checking Administration, Zhdanov dispatched his Agitprop men to investigate party administration and turn up evidence of Malenkov's negligence. Zhdanov's Agitprop apparat exposed the rival cadre apparat for ignoring the importance of ideology, neglecting party training, and allowing ideological deviations (local nationalism,

---

[76] Patolichev, *Ispytaniye*, pp. 280–84. Patolichev's first identification as CC secretary appeared in the 9 October 1946 *Pravda.*

[77] One biography—in the *Bolshaya sovetskaya entsiklopediya,* 2d ed., 1955—does indicate that Patolichev became a CC secretary in May 1946.

[78] Patolichev, *Ispytaniye,* pp. 280–84.

liberalism in culture, and so on). Later, the inspectors of the new Checking Administration took over the campaign.

The leading roles in the campaign were played by Aleksandrov (who ran Agitprop's mouthpiece, *Kultura i zhizn*), Fedoseyev (who opened Zhdanov's campaign with exposures in the Ukraine), and Iovchuk (who propagated Zhdanov's campaign at a number of conferences and who became Belorussian secretary for ideology in early 1947 when Malenkov's Belorussian protégés were purged).

Kultura i zhizn was initiated to conduct Zhdanov's ideological campaign, a fact made clear in its first issue, which editorially explained that the unsatisfactory work of many existing press organs and the need to make literature, art, films, radio, and science better serve communist education had necessitated the creation of a new newspaper. Appearing every ten days, *Kultura i zhizn* printed editorials and articles, often by Agitprop officials, which scourged various organizations and regions, as well as other press organs, for ideological shortcomings. Other papers, including *Pravda,* soon began to reprint its key editorials and articles, clearly attesting to the authoritativeness of *Kultura i zhizn* as the main vehicle of Zhdanov's crackdown.

The campaign to put new stress on ideology was itself an implicit attack on Malenkov, who had largely ignored ideology and focused the apparat on economic administration. At the 1939 party congress, Zhdanov had complained that the CC apparat was too involved in economic questions because it was organized into industrial branch sections (*otdely*) that supervised the corresponding ministries.[79] Stalin agreed,[80] and in the reorganization enacted by the congress most such sections were abolished and the CC apparat was organized largely into two big administrations (*upravleniya*)—cadres and propaganda-agitation.[81] This reorganization raised the status of

---

[79] Zhdanov declared that "production-branch sections now do not know what they really should deal with, they substitute for economic organs and compete with them, and this engenders irresponsibility in work." He proposed that they be liquidated, except for the Agriculture Section (*XVIII syezd vsesoyuznoy kommunisticheskoy partii (b)* [18th Congress of the All-Union Communist Party (Bolshevik)], stenographic report, pp. 528, 532). For background on CC organization in the 1930s, see L. A. Maleyko's article in the February 1976 *Voprosy istorii KPSS,* pp. 111–22.

[80] Stalin in his report to the congress called for concentrating ideological and cadre work in two new administrations (*XVIII syezd vsesoyuznoy kommunisticheskoy partii (b),* stenographic report, pp. 30–31).

[81] The new party statute adopted at the 1939 congress stipulated that the CC would have a Cadres Administration, Propaganda and Agitation Administration,

ideological work and restricted party interference in economic administration. This 1939 upgrading of ideological work and limitation on economic involvement by the CC apparat was gradually eroded by Malenkov as the war approached, however, and during the war the party apparat naturally concentrated its efforts on increasing production of war materials and neglected ideology.[82] Zhdanov's 1946 ideological campaign more than repaired the imbalance between ideology and economic administration within the CC apparat; by discrediting, breaking up, and reorganizing Malenkov's cadres apparat, it established the priority of ideology. Immediately upon Zhdanov's death in 1948, Malenkov reversed this stress and also won a reversal of the 1939 CC reorganization, establishing more industrial branch sections in 1948 than ever before.[83]

The negligence in cadre selection and ideological work uncovered by the Zhdanovites was blamed for allowing the growth of pressures for more freedom of expression in literature and culture and for greater latitude in which to promote local languages, cultures, and histories in non-Russian republics. Zhdanov's crackdown began in the Ukraine, where the relaxation of ideological controls during and immediately after the war had led Ukrainian historians to publish books with a less russified version of history and prompted Ukrainian writers to press for freedom from censorship and party control. These deviations were noted in Moscow, and Agitprop deputy chief Fedoseyev traveled to Kiev to oversee and address a 24–26 June conference of Ukrainian propaganda officials, writers, and teachers. Although Fedoseyev's speech was not published, the report of K. Z. Litvin, the Ukrainian secretary for propaganda, was and it clearly reflected the message from Moscow. According to the account in *Pravda Ukrainy* of 29 June, Litvin dwelt on ideological shortcomings, especially "bourgeois nationalist" errors in recent history books and the demand of some writers for "the right to make mis-

---

Organizational-Instructor Section, Agriculture Section, and Schools Section (ibid., p. 674).

[82] The swings between Zhdanov's stress on ideology and Malenkov's on economic involvement as reflected in CC organization from 1939 to 1941 are traced in detail in Jonathan Harris's "The Origins of the Conflict Between Malenkov and Zhdanov: 1939–1941," *Slavic Review*, June 1976, pp. 287–303.

[83] Including sections for light industry, heavy industry, machine building, and transport (B. A. Abramov in the March 1979 *Voprosy istorii KPSS*, p. 64).

takes," which he labeled as tantamount to demanding "the right to deviate from our Soviet ideology."[84]

Although the June conference initiated a series of local articles and meetings attacking Ukrainian nationalistic errors, Zhdanov's men in Moscow soon stepped in again. S. Kovalev, head of Agitprop's Propaganda Section, wrote an article (*Kultura i zhizn,* 20 July 1946) demanding correction of errors in the presentation of Ukrainian history. Reprinted in the 23 July *Pravda Ukrainy* and the 24 July *Radyanska Ukraina,* the article was quickly followed by more Ukrainian press attacks on deviations in history.[85]

At the same time, the CC called the Ukrainian leaders to Moscow to report on cadre work and on 26 July adopted a decree condemning their work in training and appointing leading cadres. Republic First Secretary Khrushchev spoke on this matter at a 15–17 August plenum of the Ukrainian CC, acknowledging poor selection of cadres in ideological work, and admitting that this negligence had led to a situation in which attempts to revive Ukrainian nationalism were permitted.[86] This setback to Khrushchev was also the apparent result of a check by CC inspector N. I. Gusarov on Ukrainian cadre work. According to Gusarov's son, Gusarov in 1946 gave a report criticizing Khrushchev, after which he and Khrushchev called each other names.[87] Gusarov's report was presumably the basis for his

---

[84] In the 18 August 1946 *Radyanska Ukraina,* D. Moroz revealed that two prominent Ukrainian writers, Ukrainian Writers' Union presidium members Petro Panch and Ya. Horodskoy, had recently demanded the right to make mistakes (apparently at an early June meeting of prose writers reported in the 14 June 1946 *Radyanska Ukraina*). According to Moroz, Panch had declared "give us the right to err—and our creative works will not be as boring as they often are," and Horodskoy had taken a swing at censorship, stating that "Pushkin did not clear his works with anyone." At a Ukrainian Writers' Union meeting reported in the 5 September 1946 *Pravda Ukrainy,* it was alleged that Ukrainian Writers' Union Chairman M. F. Rylskiy had backed them and was their "supporter and defender."

Among history books, the *Ocherk istorii ukrainskoy literatury* [Outline history of Ukrainian literature], written in 1942–43 and published in 1945, and the 1943 first volume of the *Istoriya Ukrainy* [History of the Ukraine] were the main targets, accused of stressing the distinctness of Ukrainian history and nationality from Russian history and nationality and of presenting some Ukrainian historical figures too favorably (see *Pravda Ukrainy* of 30 June, 2 July, 7 July, 14 July and 23 July 1946, and *Radyanska Ukraina* of 20 and 21 July 1946).

[85] See *Pravda Ukrainy* of 25 July 1946, and *Radyanska Ukraina* of 24 July, 26 July, 27 July, and 11 August 1946.

[86] *Pravda,* 23 August 1946, and *Pravda Ukrainy,* 24 August 1946.

[87] Vladimir Gusarov, *Moy papa ubil Mikhoelsa* [My papa killed Mikhoels] (Frankfurt am Main, 1978), pp. 90–92.

article in *Partiynaya zhizn* (no. 1, November 1946) criticizing Ukrainian cadre work.

A 24 August *Pravda* editorial linked the Ukrainian cadre and nationalistic errors to the ideological errors exposed by the 14 August VKP CC decree on the Leningrad journals *Zvezda* and *Leningrad*. Pointing to the August Leningrad party *aktiv* meeting (at which Zhdanov had spoken) as "an example of sharp Bolshevik criticism and self-criticism," the editorial accused the Ukrainian leadership of "underrating the special importance of ideological work," neglecting "ideological-political education of cadres of the intelligentsia," and not organizing press criticism of "hostile bourgeois nationalist ideology." It cited Panch's "theory" of a writer's right to make ideological errors (see note 84), condemning it as a call for ideological deviations.

The ideological errors in history and literature in the second largest Soviet republic were thus associated with poor cadre work—an accusation that reflected badly on the VKP's Cadres Administration and its supervisor Malenkov. Moreover, Zhdanov's campaign also weakened Khrushchev, the Ukrainian boss, who apparently was closer to Malenkov at that point.

Fedoseyev's trip to the Ukraine to expose local deviations was followed by Iovchuk's trip to Azerbaydzhan to expose ideological negligence by Beriya's close protégé, First Secretary M. D. Bagirov. A 18 July *Pravda* account of a recent Azerbaydzhan ideological conference reported that Iovchuk set forth tasks in literature, art, and social sciences and criticized local officials for neglecting Marxist-Leninist education.[88] Another Agitprop official—deputy head of Agitprop's Propaganda Section A. I. Maslin—spoke at a 15–17 July conference on Kazakh propaganda and criticized the ideological errors in history and literature recently exposed in some national republics.[89]

Exposures of ideological negligence in the Ukraine were soon extended to charges of laxity in indoctrinating the millions of new party members admitted during the war, a group that in 1946 constituted two thirds of the entire party membership. On the same

[88] Apparently about the same time, CC inspector N. I. Gusarov also attacked Bagirov. Gusarov's son mentions that his father accused Bagirov of "'un-party' behavior and other shortcomings," but was rebuffed when Bagirov appealed to Stalin for help (Vladimir Gusarov, *Moy papa ubil Mikhoelsa*, pp. 92–93).

[89] Kazakhstan got off easy, since the only Kazakh example Maslin cited was a Kazakh history book that had already been criticized in August 1945 (see the 28 July 1946 *Kazakhstanskaya pravda*).

day as the decree on the Ukraine (26 July), the CC adopted a decree noting the big influx of new members into the party in recent years and calling for more stringent admissions and better training of new members. Although the published excerpts of the decree[90] blame only local party organs, it seems that it also dealt a blow to Malenkov, who was responsible for party admissions and political work with new members.

After his 4 May removal from the Secretariat, Malenkov dropped to last place among Politburo members in protocol rankings.[91] He was apparently "on ice," having no set responsibilities, but he remained a Politburo member and continued to attend leadership functions. On 2 August he was at least given a new post—that of deputy premier—although this appointment was kept secret until mid-October.[92] He apparently hit his low point at about this time, being physically exiled from Moscow and failing to appear at the 7 November 1946 parade.[93]

Reorganization of Cadre Organs

The exposure of ideological errors at the 24–26 June Ukrainian conference and the 26 July CC decree on Ukrainian cadre errors were quickly followed by a wholesale reorganization of central cadre organs in Moscow. On 2 August 1946 the Politburo reformulated the functions of the Orgburo and Secretariat and reorganized the

[90] In *KPSS v rezolyutsiyakh i resheniyakh syezdov, konferentsiy i plenumov TsK,* vol. 6, 1971, p. 154.

[91] He was listed sixth at the Kalinin funeral (*Moskovskiy bolshevik,* 6 June 1946) and last among Politburo full members at a June Stalin dinner (*Pravda,* 9 June 1946), at the July physical culture parade (*Pravda,* 22 July), at a July dinner (*Pravda,* 26 July), at Air Day ceremonies (*Pravda,* 19 August), and at Tank Day ceremonies (*Pravda,* 9 September).

[92] The first announcement of the 2 August ukase appointing Malenkov was in a speech by A. Gorkin, secretary of the Presidium of the Supreme Soviet, which was delivered at the October Supreme Soviet session (*Pravda,* 19 October 1946). Thereafter, Malenkov was only identified as deputy premier, not as CC secretary (for example, when nominated for the RSFSR Supreme Soviet in January 1947—see *Moskovskiy bolshevik,* 4 January 1947).

[93] Khrushchev, in his recollections, indicates that Malenkov had been exiled from Moscow in 1946 (*Khrushchev Remembers,* pp. 252–53). According to Khrushchev, Malenkov's disgrace was the result of acts by Stalin's son, Vasiliy. Vasiliy, a pilot, denounced A. I. Shakhurin, minister of the aviation industry, for allowing production of defective planes. Shakhurin was thrown into prison and Malenkov, who supervised Shakhurin's ministry, was fired as CC secretary and exiled to Central Asia. On the tapes used to prepare his recollections, Khrushchev states that Malenkov "didn't remain there long, that is, and soon returned." He returned so quickly, according to Khrushchev, that "many probably don't even recall it now." He declares that "Beriya

party structure "from the CC apparat to the *rayon* party committees, inclusively."[94] This reorganization coincided with the censure of the CC organs that were supervised by Malenkov.

The reorganization of Malenkov's apparat was outlined in several decrees adopted during August. A 2 August decree, published in *Partiynaya zhizn* (no. 1, November 1946), declared that CC cadres work was poor and established a new, reorganized Higher Party School (in the Cadres Administration) and an Academy of Social Sciences (in Agitprop) to improve training.[95] Another August decree published in the 20 August 1946 *Kultura i zhizn* criticized the work of the CC journals *Partiynoye stroitelstvo, propagandist,* and *V pomoshch lektoru* as unsatisfactory and established a new journal, *Partiynaya zhizn,* to replace them. Malenkov had been and probably still was editor of *Partiynoye stroitelstvo* when it was censured and abolished.[96] The first issue of *Partiynaya zhizn* carried the article by CC inspector N. I. Gusarov criticizing Ukrainian cadre work.

---

brought him back" and notes how Beriya later bragged about how he "step-by-step, so to say, raised the question of Malenkov, that is, with Stalin, in order to get Malenkov brought back" and "eventually Malenkov returned and again occupied his post, of CC secretary, that is" (Russian-language transcript of the Khrushchev tapes, pt. I, p. 986).

Since Malenkov was present at the 8 September Tank Day ceremony, as well as at functions in June, July, and August, his exile presumably began during September or October. He had returned to Moscow by early 1947, however, appearing with the leadership at the 21 January Lenin anniversary ceremony. Khrushchev mentioned that at the end of the 21–26 February 1947 CC plenum, Malenkov was working out the plenum's decree (*Khrushchev Remembers,* p. 239).

[94] This Politburo action is mentioned in P. A. Rodionov's book *Kollektivnost: vysshiy printsip partiynogo rukovodstva* [Collectivity: a high principle of party leadership] (Moscow, 1974), p. 205. The quote is his characterization.

[95] *Pravda* announced the creation of the new Academy of Social Sciences on 30 August 1946. Zhdanov's protégé Iovchuk wrote in a February 1976 *Kommunist* (no. 3, p. 83) that these two schools were created "with the direct participation" of Zhdanov. The new schools were opened on 1 November with Zhdanov absent but his protégés or allies dominating the proceedings: CC secretaries Kuznetsov, Patolichev, and Popov; Agitprop chief Aleksandrov and deputy chiefs Fedoseyev and Iovchuk; *Pravda* editor Pospelov; and deputy chief of the Cadres Administration A. N. Larionov (*Pravda,* 2 November 1946). Agitprop chief Aleksandrov became leader of the department of history of philosophy at the new academy, according to the January 1947 *Vestnik* of the Academy of Sciences, p. 56.

According to the new CPSU history issued in 1980, a Politburo "decision" (*resheniye*) also on 2 August formally changed the Central Committee's Organizational-Instructor Section to the Administration for Checking Party Organs (*Istoriya kommunisticheskoy partii sovetskogo soyuza,* vol. 5, bk. 2, pp. 41–42).

[96] Malenkov was listed as responsible editor until late 1940, when an anonymous

Although Malenkov's deputies N. N. Shatalin, Ye. Ye. Andreyev, A. S. Pavlenko, and V. D. Nikitin apparently retained their posts in the leadership of the Cadres Administration, this organ now apparently came under Patolichev's supervision and Patolichev added his friend A. N. Larionov to its leadership. Larionov became deputy chief of the administration in September 1946, according to his 24 September 1960 *Pravda* obituary.[97] From January 1939 to December 1941 Larionov had been second secretary of Yaroslavl, while Patolichev was first secretary, and then had succeeded to Patolichev's post. Patolichev's attitude toward Larionov is clear from the considerable praise he devoted to Larionov in his autobiography.[98]

Zhdanov's hand can also be seen in Larionov's promotion, for Larionov's replacement in Yaroslavl was I. M. Turko, second secretary of Leningrad *oblast* and a protégé of Zhdanov. Malenkov's attitude toward these shifts is clear from the fact that shortly after his return to power in mid-1948, Larionov was transferred out of the Cadres Administration to become first secretary in Ryazan.[99] During the 1949 purge of Zhdanov's Leningrad protégés, Turko was persecuted personally by Malenkov, who tried to drag him into the Leningrad Case. In the 3 July 1957 *Leningradskaya pravda,* Turko related that in February 1949 Malenkov had summoned him to Moscow and with threats and accusations tried to force him to sign forged documents.

By late 1946 Patolichev was well along in staffing the new Administration for Checking Party Organs, having gathered his assistants from all over the country. G. A. Borkov apparently came to Patolichev's administration almost immediately. Released as Kazakh first secretary at a 20–22 June Kazakh plenum and placed "at the disposal" of the CC (*Kazakhstanskaya pravda,* 23 June 1946), he had in fact apparently left for Moscow weeks earlier (since he made

---

"editorial collegium" began signing off the journal. By May 1947 Malenkov's deputy, Cadres Administration deputy chief N. N. Shatalin, was listed as chief editor of the new journal *Partiynaya zhizn* (no. 10, May 1947).

[97] On 28 August 1946 *Pravda* had announced Larionov's release as Yaroslavl first secretary in connection with his assumption of "leading work" in the CC apparat, but he was first identified as deputy chief only in the 2 November 1946 *Pravda.* Neither Patolichev nor anyone else was ever identified in the press as chief of the administration.

[98] Patolichev, *Ispytaniye,* pp. 87, 94.

[99] Larionov's 24 September 1960 *Pravda* obituary indicates that this change was made in December 1948.

no appearances in Kazakhstan during June or even May). Borkov was first identified as deputy chief of the Checking Administration, however, only in the 22 January 1947 *Pravda*. S. D. Ignatyev appears to have left Bashkiria by June, but he was first identified as deputy chief of an unidentified CC administration in the 9 October 1946 *Pravda* and specifically as deputy chief of the Checking Administration in the 1 February 1947 *Pravda*. When N. M. Pegov left Primorskiy *kray* is unclear but he was first identified as deputy chief of the Checking Administration in the 21 May 1947 *Kultura i zhizn*. Perm First Secretary N. I. Gusarov, according to his son's book, was appointed a CC inspector in the spring of 1946.[100] Kemerovo First Secretary S. B. Zadionchenko was first identified as a CC inspector in the 29 September 1946 *Pravda*. The apparent first deputy head, V. M. Andrianov, was never publicly identified with the administration.

The new administration was in operation by August. *Partiynaya zhizn* (no. 1, November 1946, signed to press 15 November 1946), reported that a responsible organizer of the Checking Administration named Vakulenko had reported to the CC on the work of a Rostov *rayon* party committee and that on the basis of this report the CC on 17 August 1946 had adopted a decree criticizing the *rayon* committee. This was apparently the first press reference to the new administration.

At the end of 1946 the new Administration for Checking Party Organs was unveiled and explanations of its purpose began to appear in CC publications. L. Slepov, head of *Pravda*'s department for party life,[101] announced in the issue for 14 December 1946 that the Organizational-Instructor Section had recently been reorganized into an Administration for Checking Party Organs and explained that the new administration was charged with inspecting the work of *oblast* party committees, *kray* party committees, and republic CCs and their fulfillment of CC decrees. He noted that a "group of inspectors" was being formed in this administration consisting of "authoritative" and "experienced" officials who could "give orders locally" to improve the work of party organizations. A. Shvarev,

---

[100] Vladimir Gusarov, *Moy papa ubil Mikhoelsa*, p. 90.

[101] Slepov was so identified in the 11 April 1948 *Kultura i zhizn*. According to his 27 October 1978 obituary in *Moskovskaya pravda*, Slepov had been deputy head of the Organizational-Instructor Section during the war (probably until its abolition in 1946).

himself later identified as a CC inspector,[102] wrote about checking and the creation of the new administration in a February 1947 *Partiynaya zhizn* (no. 3). He stated that "the significant renewal of leading party cadres occurring during the war" (that is, the large number of new officials appointed under Malenkov's stewardship) had created a need for new measures to ensure correct implementation of decisions. "For this purpose," wrote Shvarev, "the institution of inspectors was created in the CC" to check more closely on local officials.

Having been given considerable authority, the new inspectors proceeded to tackle prominent regional leaders. Gusarov's son relates how his father proudly described his new position as "personal representative of Stalin" and considered his authority great enough to attack Ukrainian First Secretary Khrushchev and Azerbaydzhan First Secretary Bagirov. Gusarov's son describes his father and Khrushchev calling each other names at a 1946 conference after a report by Gusarov criticized Khrushchev.[103] This report was presumably what led to the 26 July CC decree condemning cadre work in the Ukraine; it also apparently served as the basis for Gusarov's article in *Partiynaya zhizn* (November, 1946) criticizing Ukrainian cadre work. Having successfully contested with Khrushchev, Gusarov then started criticizing Beriya's protégé Bagirov for "unparty" behavior and other shortcomings. But in this case, Bagirov called Stalin and got Gusarov slapped down.[104]

While inspector Gusarov took on Khrushchev and Bagirov, inspector S. B. Zadionchenko tackled P. K. Ponomarenko, Belorussian first secretary and premier, whose closeness to Malenkov made him an especially obvious target for Zhdanov. He had been Malenkov's deputy in the CC's Leading Party Organs Section (the earlier name for the cadres department) before becoming Belorussian first secretary in 1938.[105] Zadionchenko conducted an investigation of Belorussia's affairs in late 1946 and apparently presented a negative picture of the first secretary's work. On 25 January 1947 Ponomarenko and Zadionchenko both delivered reports to the CC on the Belorussian CC's work, and the CC adopted a decree censur-

---

[102] Shvarev was identified as a CC inspector in the 23 March 1949 CC decree on Gorkiy *oblast* (*KPSS v rezolyutsiyakh i resheniyakh syezdov, konferentsiy i plenumov TsK*, vol. 6, 1971, p. 277).
[103] Vladimir Gusarov, *Moy papa ubil Mikhoelsa*, pp. 90–92.
[104] Ibid., pp. 92–93.
[105] See his biography in the 1 March 1950 *Sovetskaya Belorussiya*.

ing the Belorussian party for failures in leadership of industrial, agricultural, and ideological work and criticizing Ponomarenko personally for "incorrectly organizing the work" of the Belorussian CC. It was alleged that the Belorussian leaders had neglected literature, art, science, and culture, allowed "nonideological" literary works to appear, and failed to take action after the CC decree on *Zvezda* and *Leningrad*.[106] Zadionchenko also wrote up Belorussia's errors in *Partiynaya zhizn* (no. 2, 1947).

On 7 March 1947, Ponomarenko and most other top Belorussian leaders were replaced by a team from Moscow. CC inspector N. I. Gusarov was elected Belorussian first secretary. S. D. Ignatyev, the deputy chief of the CC's Administration for Checking Party Organs, became Belorussian secretary for agriculture. Zhdanov's protégé M. T. Iovchuk, the deputy chief of Agitprop, became Belorussian propaganda secretary, while Ponomarenko retained his lesser post as republic premier.

The Belorussian purge was reversed as soon as Zhdanov faded and Malenkov regained power. In July 1948 Ponomarenko was called to Moscow to become a member of the VKP CC Secretariat and Orgburo,[107] and during 1949–50 all three leaders who had been sent to Belorussia in March 1947 were demoted. Ponomarenko was given the satisfaction of returning to Belorussia in 1950 on a special assignment to expose errors by Gusarov, after which Gusarov was removed as republic first secretary in July 1950. Gusarov was again appointed CC inspector, but, as his son points out, this time the appointment was a demotion.[108] Iovchuk was packed off to study at the Academy of Social Sciences,[109] while Ignatyev was made a CC

---

[106] The decree on Belorussia apparently was never published in the press. However, it was printed in the 1971 *KPSS v rezolyutsiyakh i resheniyakh syezdov, konferentsiy i plenumov TsK*, vol. 6, p. 195.

[107] The exact date of Ponomarenko's appointment to the Secretariat has never been made public, but it was apparently in July 1948. He was last identified as being in Belorussia by the 27 June 1948 *Sovetskaya Belorussiya*, and on 25 July 1948 for the first time he stood with VKP Politburo members and secretaries at a parade in Moscow (*Pravda*, 26 July 1948). He was relieved as Belorussian premier on 12 September 1948 (*Sovetskaya Belorussiya*, 14 September 1948). Ponomarenko's biography in the 1 March 1950 *Sovetskaya Belorussiya* listed him as VKP CC secretary starting in 1948.

[108] Ponomarenko's 1950 mission to Belorussia to demote his successor Gusarov is described in Vladimir Gusarov's *Moy papa ubil Mikhoelsa*, pp. 95, 103.

[109] According to his biography in the *Bolshaya sovetskaya entsiklopediya* yearbook for 1971.

plenipotentiary for Uzbekistan (he is so identified in the 24 February 1950 *Pravda*).

The inspectors also played important roles in other, later shake-ups of powerful local leaders. On 28 February 1948 Zadionchenko and A. I. Struyev, first secretary of Stalino *oblast,* delivered joint reports to the CC on the situation in the *oblast,* which led to a decree criticizing the Stalino *oblast* party committee's management of the coal mines.[110] On 23 March 1949 inspector Shvarev, along with Gorkiy First Secretary S. Ya. Kireyev, reported to the CC on the Gorkiy situation, and the CC adopted a decree criticizing "big errors" in the work of the Gorkiy party.[111] Kireyev was removed soon afterward.

The Crackdown

The exposures of deviations and laxness in the Ukraine and elsewhere and the reorganization of Malenkov's apparat were followed by a crackdown that developed along several lines. The best known was the drive for reassertion of ideology, traditionally known as the *Zhdanovshchina,* which began in August. Then in September came the campaign against violations by farmers, notably the expansion of private farming at the expense of public farming, and CC decrees censuring various oblasts for nonfulfillment of agricultural procurement quotas. Although the crackdown clearly began in August, the detailed public announcements—Zhdanov's speeches and the decree on agricultural abuses—appeared in print only later in September.

This public crackdown, especially in culture, is the basis of Zhdanov's reputation as a hard-line ideologue. He indeed made vicious statements during this campaign, but even such an inside observer as Khrushchev has suggested that Zhdanov's reputation is not fully deserved. Khrushchev—a prime victim of Zhdanov's 1946 campaign and surely no friend—declared in his taped recollections that he later realized that the intelligentsia still harbored strong dislike for Zhdanov because of his postwar speeches on culture and his suppression of the Leningrad journals. But, Khrushchev states,

---

[110] *KPSS v rezolyutsiyakh i resheniyakh syezdov, konferentsiy i plenumov TsK,* vol. 6, 1971, p. 261.

[111] Ibid., p. 277. This decree, however, was part of the purge of Zhdanov's men in his former bailiwick of Gorkiy. By then, the inspectors were being used as hatchet men for Zhdanov's foes.

while "Zhdanov played his individual role" in this campaign, "he was still mainly carrying out Stalin's direct instructions," and "I think that if Zhdanov himself had set his own attitude and his own policy in these questions it would not have been so rigid, because you really cannot regulate the development of literature, the development of art, the development of culture with a stick, with shouting."[112]

The *Zhdanovshchina* in literature began in Zhdanov's own Leningrad, prompted by a 26 June decision of the Leningrad city party committee which confirmed the new editorial board of the Leningrad literary journal *Zvezda.* Included on the new board was satirist Mikhail Zoshchenko, who had gotten into trouble a couple of years earlier for irreverent writings, but who was again being accepted in the more tolerant postwar atmosphere.

On 20 August *Kultura i zhizn* announced a 14 August 1946 CC decree that criticized the journals *Zvezda* and *Leningrad* for publishing "ideologically harmful" works, especially the "hostile" writings of Zoshchenko and Anna Akhmatova, the controversial poetess. The decree condemned as a "gross political error" the 26 June decision of the city party committee to add Zoshchenko to *Zvezda*'s editorial collegium, abolished the journal *Leningrad,* and named Agitprop deputy chief A. M. Yegolin new chief editor of *Zvezda.* The CC decree also criticized Agitprop and the USSR Writers' Union and its chairman N. S. Tikhonov for poor supervision of the two journals. *Pravda* printed excerpts from the decree on 21 August. On 22 August the same newspaper revealed that Zhdanov had spoken on the 14 August decree at a recent meeting of the Leningrad party *aktiv* and at a Leningrad Writers' Union meeting, but it neither carried excerpts from Zhdanov's speeches nor indicated the dates of the meetings.[113]

The campaign against ideological errors quickly picked up speed, and many of the shortcomings cited were those exposed in the Ukraine by Zhdanov's Agitprop men. On 23 August *Pravda* carried a long account of the 15–17 August Ukrainian CC plenum on cadre errors, and a 24 August *Pravda* editorial praised Zhdanov's Lenin-

---

[112] Russian-language transcripts of the Khrushchev tapes, pt. I, p. 872.

[113] The 1957 volume issued on Leningrad's anniversary (*Leningrad: entsiklopedicheskiy spravochnik,* p. 514) stated that the party *aktiv* meeting was on 15 August 1946 and the Writers' Union meeting on 16 August 1946.

grad speech and criticized the Ukraine for ideological errors. On 24 and 29 August *Pravda* attacked the Ukrainian publications *Perets* and *Radyanska Ukraina.*

New CC decrees on ideology, *Pravda* editorials, and *Kultura i zhizn* attacks followed. A 23 August *Pravda* editorial revealed that "recently" the CC had adopted a decree on improving cadre training and creating an Academy of Social Sciences and Higher Party School. On 30 August *Kultura i zhizn* announced a 26 August CC decree on ideological shortcomings in theater repertoires, and *Pravda* also discussed the same decree on 2 September. On 8 September *Pravda* reported that a 31 August– 4 September meeting of the Writers' Union presidium had discussed the 14 August CC decree, removed Tikhonov as the union's chairman, and expelled Zoshchenko and Akhmatova. The same meeting also assailed Panch's "theory" of a "writer's right to err." A 11 September *Pravda* article on cinema revealed that on 4 September the CC had adopted a decree criticizing the film "Great Life." Finally, on 21 September *Pravda* published the long report on *Zvezda* and *Leningrad* which Zhdanov had delivered at the Leningrad party *aktiv* and Writers' Union meetings in August. (*Kultura i zhizn,* which came out only every ten days, carried the speech in its issue for 30 September.) The decrees on the Leningrad journals, theater repertoires, and the film "Great Life" constituted the core of Zhdanov's ideological campaign.

The tightening up in agricultural policy began when the CC and Council of Ministers issued a joint decree (19 September 1946) which condemned abuses by kolkhozes, especially their allowing *kolkhozniks* illegally to expand their private plots, and established a Council for Kolkhoz Affairs to supervise kolkhozes and prevent such abuses (*Pravda,* 20 September 1946). The decree was signed by Stalin and Zhdanov, and signaled the start of a new campaign. When the membership of the new council was announced in the 9 October *Pravda,* it could be seen that, like the new CC Checking Administration, it was to be an instrument for tightening controls. The council was authorized to have checkers (*kontrolery*) who could monitor local authorities, and in fact three of the new council's key officials were leaders of the Checking Administration—a fact that was carefully hidden at the time. Patolichev, chief of the Checking Administration, and first deputy chief Andrianov were named the two deputy chairmen of the new council, while deputy chief Ignatyev was also among its thirty-nine members. However, *Pravda* only identified

Patolichev as CC secretary, Andrianov as an Orgburo member, and Ignatyev as deputy chief of an unnamed CC administration.

The creation of the new council was part of a general crackdown on *oblast* authorities which focused on agricultural shortcomings and was occasioned by the agricultural disaster developing during the second half of 1946. A severe drought set in during mid-1946 and by August there was widespread fear of famine.[114] Stalin put heavy pressure on local authorities to reinforce discipline and to extract as much grain as possible from the farms. This pressure was reflected in sharp CC criticism of the agricultural leadership in a number of *oblasts* and the removal of several *oblast* leaders.[115]

The Heir Apparent

During the crackdown, Zhdanov was clearly established as Stalin's heir apparent in the party. Some of his Leningrad protégés assumed the leadership of other provinces and his allies Andreyev and Voznesenskiy were promoted to new positions of power, while

[114] Alexander Werth, who toured parts of the south in August 1946, found people fearful of hunger in the coming months. For more on the drought and resulting famine, see pages 148–49 and 152–53 of his *Russia: The Post-War Years.* Finance Minister A. G. Zverev, in his speech to the 20 February 1947 session of the Supreme Soviet, called the 1946 drought the worst in fifty years (*Pravda,* 21 February 1947). Malenkov called it "one of the worst droughts in the history of our country" and indicated that it had forced postponement of the end of rationing (*Pravda,* 9 December 1947). Khrushchev devoted much attention to the famine of 1946–47 in his recollections, describing mass starvation and even some cannibalism (*Khrushchev Remembers,* pp. 228–35). When in power, he also had made public references to starvation, declaring in his report to the December 1963 plenum of the CC: "Yes, comrades, this is a fact, that in 1947 in a number of *oblasts* of the country, for example, in Kursk, people died of hunger" (*Pravda,* 10 December 1963). The grain harvest in 1946 fell to 39,600,000 tons as against 47,300,000 in 1945 and 95,600,000 in 1940 (*Istoriya kommunisticheskoy partii sovetskogo soyuza,* vol. 5, bk. 2, pp. 174–75).

[115] Concern over the threat of shortages was evidenced by the Politburo's 27 September 1946 discussion of food reserves and ways to save grain (mentioned in the *Istoriya kommunisticheskoy partii sovetskogo soyuza,* vol. 5, bk. 2, p. 25). In addition, a September 1946 CC decree criticized the Ivanovo *oblast* leadership (*Partiynaya zhizn,* no. 3, February 1947), and the 9 October 1946 *Pravda* editorial on fulfilling grain procurement singled out Ivanovo for criticism. Before long, G. M. Kapranov, first secretary of the *oblast* party committee, was demoted to deputy minister of light industry for the RSFSR and replaced by V. V. Lukyanov, the first secretary of the Kirov party. A November 1946 editorial in *Partiynaya zhizn* (no. 2) revealed that Stavropol First Secretary A. L. Orlov had been fired by the CC for not pressing hard enough to force fulfillment of grain procurement quotas (Orlov's 5 July 1969 obituary in *Pravda* indicated that he had then been sent to study in the Higher Party School). The 27 September 1946 *Pravda* reported plenums (Georgia and Tambov) discussing local violations in kolkhozes, and a February 1947 *Partiynaya*

those more closely identified with opposing factions—Yudin, Ponomarenko, Khrushchev—suffered sharp setbacks. By early 1947, Zhdanov had reached the peak of his power.

As the crackdown unfolded, Zhdanov's personal status rose sharply. On 20 September *Pravda* published a decree on the kolkhoz system signed by Premier Stalin on behalf of the Council of Ministers and by CC Secretary Zhdanov on behalf of the party—a clear signal that Zhdanov was now Stalin's top deputy. The next day, Zhdanov's Leningrad speech of August filled almost two pages in *Pravda,* and on 6 November it was Zhdanov who delivered the October Revolution anniversary speech. The fact that Zhdanov exposed ideological errors made by his protégés in Leningrad (including his appointees as city party secretaries), instead of reflecting badly on him, redounded to his credit: the 24 August *Pravda* editorial held up his action as "an example of sharp Bolshevik criticism and self-criticism."

Zhdanov's growing influence was also evident in the promotion of some of his Leningrad protégés. On 15 August *Pravda* reported a plenum of the Crimean *oblast* party committee at which CC representative N. Ya. Itskov criticized *oblast* First Secretary P. F. Tyulyayev and Second Secretary V. A. Berezkin, both of whom were subsequently removed. The new *oblast* first secretary was N. V. Solovyev, chairman of the executive committee of Leningrad *oblast.* Khrushchev later indicated that he regarded the Crimean party organization as having been taken over by Leningraders and he stated that these Crimean leaders were arrested along with Zhdanov's men in Leningrad and Gorkiy in 1949–50: "I recall that

---

zhizn (no. 3) indicated that the CC had found shortcomings in adherence to kolkhoz statutes in Kuybyshev, Tambov, Chkalov, and elsewhere. In January 1947 *Partiynaya zhizn* (no. 1) revealed that the CC had recently adopted decrees criticizing the Saratov and Tambov party organizations for poor preparations for the spring sowing. On 22 January 1947, *Pravda* reported a Saratov plenum at which Borkov, deputy chief of the CC Checking Administration, spoke about the CC decree on Saratov and criticized *oblast* First Secretary P. T. Komarov and others for lagging in agricultural work. Four days later *Pravda* reported a Tambov plenum at which Nikitin, the deputy chief of the Cadres Administration, discussed the CC decree on Tambov and criticized local agricultural shortcomings. On 29 January 1947, *Pravda* reported a Kursk plenum on a CC decree that had criticized local preparations for the 1947 sowing. Vladimir First Secretary G. N. Paltsev was removed around the end of 1946, transferred to economic work, and replaced by Cadres Administration official P. N. Alferov (see Alferov's 14 March 1971 obituary in *Pravda*). On 19 December 1946, *Pravda* reported the removal of Stalingrad First Secretary A. S. Chuyanov. According to Chuyanov's obituary in the 3 December 1977 *Pravda,* he thereafter was engaged in "state and economic work," obviously a demotion.

in the Crimea the leadership was formed from leaders of Leningrad and that they were also arrested [in the Leningrad purge]."[116] Leningraders also took over the Yaroslavl organization: on 28 August *Pravda* reported that Yaroslavl First Secretary Larionov, who was being transferred to Moscow (as new deputy chief of the CC Cadres Administration), was being replaced as *oblast* first secretary by I. M. Turko, second secretary of Leningrad *oblast*.

At this same time, a prominent figure in the ideological field who was an open enemy of Zhdanov's protégés in philosophy was ousted. On 30 October *Kultura i zhizn* revealed that P. F. Yudin, chief of OGIZ (the Association of State Publishing Houses of the RSFSR), had been called to the CC to account for his work and subsequently removed. Yudin's leadership of OGIZ already had been sharply attacked in the second issue of *Kultura i zhizn* (10 July), and the 30 October account called "former chief" Yudin's leadership irresponsible. On 20 November the newspaper indicated that Yudin was now being made chief editor of *Sovetskaya kniga*. Later, Yudin appears to have played a key role in the intrigues that undermined Zhdanov himself (see Chapter 2).

The new Kolkhoz Affairs Council also represented a clear gain for Zhdanov and a setback for Malenkov. Malenkov, after being removed as CC secretary in May 1946, had finally been given new responsibilities on 2 August when he was named deputy premier. In this post, he was assigned to agriculture,[117] sharing this field with Andreyev, who had been dropped from the post of CC secretary in March 1946 and named deputy premier for agriculture. Some of the ministers in the agricultural sphere (including A. I. Kozlov, minister for livestock raising[118]) apparently looked to Malenkov and others

---

[116] Russian-language transcripts of the Khrushchev tapes, pt. I, p. 990.

[117] According to Khrushchev (*Khrushchev Remembers*, p. 236), Malenkov was in charge of agriculture in late 1946 and early 1947.

[118] Kozlov had apparently headed a sector in the Organizational-Instructor Section (an A. Kozlov was so identified in the January 1946 *Partiynoye stroitelstvo*) before being named head of the new Ministry for Livestock Raising, which was created on 26 March 1946. His appointment to the ministry came while Malenkov was still politically strong. In 1947, as Malenkov's authority began rising again, Kozlov became head of the CC Agriculture Section under Malenkov and, according to Khrushchev, worked with Malenkov to undermine Khrushchev. As Khrushchev related in a 2 November 1961 speech (Khrushchev's speeches, vol. 6, pp. 57–59), Malenkov and Kozlov had persuaded Lysenko to write an article opposing the sowing of winter wheat after the February 1947 plenum had censured Khrushchev for supporting winter wheat against spring wheat. When Malenkov became premier in March 1953, he named Kozlov his agriculture minister. Kozlov was transferred to

(including the procurement minister, B. A. Dvinskiy[119]) to Andreyev. The balance between Andreyev and Malenkov was upset when the Kolkhoz Affairs Council was created and Andreyev named its chairman. Andreyev, according to his biography in the 1950 *Bolshaya sovetskaya entsiklopediya* (2d ed.), had even participated in drafting the 19 September decree that created the council. His protégé Patolichev became its deputy chairman, whereas Malenkov and his protégé Kozlov were not even included. Subsequently, Andreyev served as main spokesman on agriculture, for example, delivering the main report at the February 1947 plenum on agriculture.[120] Malenkov's role had apparently become more important by that time, but was not equal to Andreyev's.[121]

Zhdanov's side scored another gain, when in October 1946 Stalin reorganized the Politburo, bringing Zhdanov's former protégé N. A.

---

minister of state farms in September 1953 and apparently followed Malenkov's line in resisting Khrushchev's 1954 plan for organizing more state farms in the virgin lands. (At the December 1958 plenum [*Pravda*, 19 December 1958], K. G. Pysin, first secretary of Altay *kray*, told how Malenkov, then premier, had signed a decree in mid-1954 which forbid Altay leaders to create new virgin-land state farms—a move obstructing Khrushchev's virgin-land campaign.) Khrushchev attacked Kozlov during 1954 and 1955, and he was ousted as minister in March 1955, immediately after Malenkov resigned as premier and confessed to errors in agricultural leadership.

[119] Dvinskiy had been appointed in 1944 (according to his obituary in the 11 June 1973 *Pravda*), while Andreyev was CC secretary in charge of agriculture and also people's commissar for farming. He was replaced with Malenkov protégé P. K. Ponomarenko in 1950.

[120] Khrushchev relates how Stalin proposed holding a CC plenum on agriculture in 1947 and decided that Andreyev should make the main report rather than Malenkov. As Khrushchev tells it, Stalin declared that even though Malenkov was in charge of agriculture, "what kind of report can he deliver," since "he doesn't know the first thing about agriculture" (*Khrushchev Remembers*, p. 236).

[121] Khrushchev (ibid., p. 239) stated that Andreyev and Malenkov had been assigned to work out the draft of the February 1947 plenum's decree. Regardless of the nature of earlier relations between Malenkov and Andreyev, their positions as competing deputy premiers for agriculture surely turned them into rivals—as subsequent events showed. Malenkov's influence grew during 1947 and he soon achieved the upper hand. When nominated for the Moscow soviet in November 1947, Malenkov was given credit for the good 1947 harvest. The official from the agriculture ministry who nominated Malenkov said that "this outstanding success was achieved with the direct participation and leadership of comrade Malenkov as deputy chairman of the USSR Council of Ministers" (*Pravda*, 23 November 1947). With the purge of Zhdanov's men following Zhdanov's death, Andreyev clearly fell into disfavor. In February 1950 he was forced publicly to confess error in having advocated the "link" (zveno) system of farming (*Pravda*, 28 February 1950), and when the Politburo was replaced by a Presidium at the October 1952 party congress, Andreyev was the only Politburo member not included in the new Presidium. By 1952 Malenkov had clearly succeeded Andreyev as Politburo spokesman on agriculture and

Voznesenskiy into its inner councils.[122] In his 1956 secret speech Khrushchev noted that on 3 October 1946 Stalin had given the six-member Politburo commission for foreign affairs responsibility for domestic affairs as well, and had added Voznesenskiy to its membership.[123] Voznesenskiy was formally promoted from Politburo candidate member to full member at a February 1947 CC plenum.

The Zhdanov faction scored further gains at this 21–26 February 1947 plenum, which heard Andreyev report on agriculture and adopted decisions devastating to Ukrainian First Secretary Khrushchev and Belorussian First Secretary Ponomarenko.[124] Khrushchev later described how, during a plenum intermission, Stalin had reproached him for planting winter instead of spring wheat in the Ukraine and how he had argued with Stalin over this.[125] The plenum's decree, published in the 28 February 1947 *Pravda*, ordered

in 1953 he abolished Andreyev's Kolkhoz Affairs Council. Khrushchev later claimed that the council had been dissolved only because of Malenkov's dislike for Andreyev. (See his 31 March 1961 Note to the Presidium, in Khrushchev's speeches, vol. 5, p. 349).

In the same note, Khrushchev declared that Malenkov had been the one responsible for agriculture at the time the council was abolished. In resigning as premier in early 1955, Malenkov acknowledged that he had been responsible for supervising agricultural organs for the previous several years (*Pravda*, 9 February 1955). The council was quietly abolished in 1953 (see the note to the 8 October 1946 decree announcing the members of the Kolkhoz Affairs Council in *KPSS v rezolyutsiyakh i resheniyakh syezdov, konferentsiy i plenumov TsK*, vol. 6, 1971, p. 194). *Pravda*'s 7 December 1971 obituary of Andreyev lists him as chairman of the council, 1946–53. Although the council continued to exist until 1953, it was rarely mentioned in its later years and probably was inactive.

[122] Voznesenskiy had gone to Leningrad in 1935 to become chief of the city planning commission and deputy chairman of the city soviet under Zhdanov. Obviously enjoying Zhdanov's patronage, he quickly became chairman of Gosplan for the whole USSR in 1938, at the age of only thirty-five. In 1939 he also became USSR deputy premier and in 1941 a Politburo candidate member. Following Zhdanov's death, Voznesenskiy was removed from his state posts in March 1949, his writings were condemned in a 13 July 1949 CC decree "On the Journal *Bolshevik*," and he was executed in 1950 (the *Malaya sovetskaya entsiklopediya* gives his date of death as 30 September 1950).

[123] Khrushchev's secret speech, p. S62.

[124] The dates of the plenum, 21–26 February, are apparently first given only in the 1971 collection, *KPSS v rezolyutsiyakh i resheniyakh syezdov, konferentsiy i plenumov TsK*, p. 210.

[125] He described Stalin's rebukes to him in his 2 November and 26 November 1961 speeches (Khrushchev's speeches, vol. 6, pp. 57, 177–79). Khrushchev also recalls that when Stalin, during a plenum intermission, had asked him what he thought of Andreyev's report, he had called it a "second-rate job," which angered Stalin (*Khrushchev Remembers*, p. 237).

expansion of spring wheat in the Ukraine by 182,000 hectares in 1947, despite Khrushchev's opposition.[126] Khrushchev later explained that the decree had been aimed at him personally: "It is true that my name was not mentioned. But anyone knowing even a little about the essence of the matter understood that this was aimed against me personally."[127] At the conclusion of the plenum, the Politburo ordered that the posts of first secretary and premier no longer be held by one person in either the Ukraine or Belorussia—an order aimed at Khrushchev and Ponomarenko.[128] As noted above, Ponomarenko lost his post as Belorussian first secretary on 7 March 1947, and a team of CC officials from Moscow took over Belorussian CC leadership. Similarly, a 3 March 1947 Ukrainian CC plenum removed Khrushchev as first secretary and replaced him with L. M. Kaganovich, a deputy premier of the USSR. CC Secretary Patolichev came down to Kiev with Kaganovich to become Ukrainian agriculture secretary, and together they forced the Ukraine to expand spring wheat sowing. Khrushchev was left with the secondary post of premier, and in late March he also lost the posts of first secretary of Kiev *oblast* and city. Khrushchev's career seemed at an end, and he soon dropped out of sight.[129]

By early 1947 so many Zhdanov allies or protégés had been sent to replace Malenkov men that several CC posts fell vacant. There is no indication that anyone headed either the CC Cadres Administration or the Checking Administration in early 1947, although deputy chiefs Shatalin[130] and Andrianov may have run them. Zhdanov's protégé A. A. Kuznetsov appears to be the only CC secretary who might have supervised cadre organs, although Khrushchev said that Kuznetsov had been put in charge of state security at some unspecified point during this period.[131] Suslov, who became a CC secretary later in 1947, was engaged in unspecified CC work during 1946 and

---

[126] Khrushchev states that he told Stalin that he opposed spring wheat for the Ukraine and that if the plenum resolution ordered expansion of spring wheat in the Ukraine, he should be recorded in opposition (*Khrushchev Remembers*, p. 239).

[127] Khrushchev's 2 November 1961 speech, in Khrushchev's speeches, vol. 6, p. 57.

[128] This 27 February 1947 Politburo decision is revealed in P. A. Rodionov's 1974 book, *Kollektivnost: vysshiy printsip partiynogo rukovodstva*, p. 205.

[129] In *Khrushchev Remembers* (pp. 242–43), Khrushchev declares that he had caught pneumonia at this time and was in bed for over two·months. In any case, he reappeared in September after a four-month disappearance, and on 26 December 1947 he was restored to the post of republic first secretary.

[130] However, Shatalin was identified as chief editor of *Partiynaya zhizn* and as a CC inspector in the 4 December 1947 *Vechernyaya Moskva*.

[131] Khrushchev's secret speech, p. S45.

we must presume that he focused on ideological work, for
ded Aleksandrov as Agitprop chief in late 1947.

by early 1947 Zhdanov appeared at the peak of his power.
He was not only in clear control of the ideological apparatus but had
neutralized the cadres apparatus, which had been the stronghold of
his leading rival. He had been put in charge of Stalin's whole effort to
crack down on deviations and laxness and was spearheading a
well-publicized campaign to restore ideological orthodoxy. Clearly,
Zhdanov had become Stalin's top deputy and heir apparent.

In reaching this position, however, Zhdanov had identified himself with harsh acts and orthodox ideological concepts, giving rise to
the hard-line image that has persisted to this day. This image has
largely obscured Zhdanov's links to more moderate forces and his
more restrained statements of early 1946. As the events recorded in
this chapter suggest, however, Zhdanov's hard-line image developed
out of his protracted struggle against Malenkov for priority in the
leadership and was a natural result of the political positions in which
the two antagonists found themselves.

CHAPTER 2

The Great Debates — 1947

The relatively free discussion of ideas, including some that were unorthodox, which characterized the immediate postwar period came to an end in late 1947 and early 1948, coincident with the development of the cold war and the division of the world into "two camps." Three great debates—in philosophy, biology, and economics— occurred during 1947 as conservatives challenged the prevailing more moderate views held by the establishment in these three fields. Initially the debates were scholarly, unfettered discussions between opposing sides, but by late 1947 and early 1948, as it became clear that Stalin had chosen to support the conservative position, the debates became a one-sided flogging of erroneous moderate viewpoints.

Although Zhdanov began 1947 in a very strong political position, the bases for his downfall began to develop in these debates. He supported the losing, more moderate side in the philosophy and biology debates and, though there is no evidence that he was involved at all in the debate on economic science, the imposition of rigid views in that field probably did not help him. In addition to being caught off guard by Stalin's support of dogmatists in ideology and social sciences, Zhdanov also lost some ground in personnel changes when Malenkov returned to the Moscow leadership and Suslov, a conservative ideologue, took over Agitprop from Zhdanov's protégés.

The three debates were roughly parallel: a group of nationalistic, aggressively ambitious scholar-politicians laid down their rigid version of "orthodoxy" and attacked established scholars and officials who had tended to downplay narrow political restraints on scientific research and thought, who admired Western science, and who expected postwar cooperation with the West. The adoption of the "two camps" theory in late 1947 naturally put these moderates at some disadvantage because of their less rigid attitude toward the West. The ambitious dogmatists, having won Stalin's support, eventually prevailed and seized many of the leading posts in these fields. In philosophy, the struggle was between the dogmatists P. F. Yudin, M. B. Mitin, and A. A. Maksimov, and Zhdanov's clique of younger, apparently more moderate Agitprop and philosophy officials—the very men who, with Zhdanov's help, had pushed the dogmatists aside since 1944. The debate in biology ranged T. D. Lysenko against the supporters of Western theories of genetics. These two debates were considerably intertwined, and the most outspoken foe of dogmatism in both fields was the philosopher B. M. Kedrov. In economic science, the debate centered on the conservatives' attacks on the positions of economist Ye. S. Varga, whose pragmatic, realistic view of Western capitalism conflicted with the "two camps" theory.

The first sign of trouble occurred as early as late 1946 or early 1947, when Stalin surprised Zhdanov and his Agitprop clique by declaring that Zhdanov's top deputy, Aleksandrov, had made major ideological errors in a book on Western philosophy. Zhdanov's clique at first tried to soft-pedal this issue, holding an unpublicized discussion in January 1947 to criticize the book. When the anti-Zhdanov dogmatists labeled this session a cover-up, however, a well-publicized debate was held in June 1947, at which Zhdanov and others roundly condemned Aleksandrov for errors. The net result was a significant setback for Zhdanov: Aleksandrov was removed as Agitprop chief and replaced by new leaders not linked with Zhdanov. Zhdanov's monopoly in Agitprop began to break up.

Although Stalin's charges against Aleksandrov were not made public and remain somewhat unclear, perhaps the most important theme in other attacks was that he had neglected or underrated Russian scientific achievements. This same theme constituted the main public charge against Zhdanov's followers in philosophy as soon as Zhdanov had been defeated in mid-1948, and it formed the basis for the anticosmopolitanism campaign that brought the purge of his clique of philosophers and ideologists in late 1948 and early

1949. This theme also coincided with the issues that underlay the other great debate of 1947, that in science, especially biology. Zhdanov and his followers placed considerable faith in Western scientific ideas and fought the views of Lysenko, Russia's home-grown scientist. After Zhdanov's death, his protégés were accused of cosmopolitanism and antipatriotism because, it was alleged, they preferred the ideas of Western geneticists to those of Lysenko, and they overrated the work of Western philosophers. These accusations were made during the campaign to claim Russian authorship of such inventions as radio, the airplane, the steam engine, and so on. Thus, although Zhdanov was narrow and dogmatic in his pronouncements on literature, music, and other spheres of culture, he resisted the dogmatists in other fields under his supervision, namely science and philosophy. His foes took advantage of this relatively moderate position to discredit Zhdanov in Stalin's eyes.

Zhdanov's moderate bent in science and philosophy may have been the result of his son's influence. By late 1947, Zhdanov had lost most of his control over Agitprop, his erstwhile bailiwick, but sometime during 1947 his son became head of Agitprop's Science Section and, as such, openly supported the anti-Lysenkoists in biology. Moreover, as head of this section he also must have sanctioned the activities of *Voprosy filosofii* editor Kedrov and other liberal philosophers. Kedrov, far from being discouraged by the 1947 attacks on Aleksandrov, pressed for further relaxation of ideological constraints on philosophical thought.

The dogmatists, after attacking Kedrov, Aleksandrov, and Fedoseyev during the June philosophy debate, pressed their attack in late 1947. Their main vehicle was, oddly enough, the organ of the Writers' Union, *Literaturnaya gazeta*. V. V. Yermilov, the conserva-tive chief editor of *Literaturnaya gazeta,* and science editor M. B. Mitin defended Lysenko and conducted a strenuous campaign against Kedrov. Though it is not clear who was behind them, the fact that the conservative M. A. Suslov had now replaced Aleksandrov as chief of Agitprop—the organ that supervised the Writers' Union— may have played an important role in this duel.

In mid-1948, Zhdanov's foes Malenkov, Beriya and Suslov, and the dogmatists in science and philosophy prevailed, as Stalin took their side against Zhdanov. Zhdanov's son, Yuriy, was forced to write a 10 July 1948 letter confessing that he had wrongly supported Lysenko's foes; at an August session of VASKhNIL (the Lenin

Agricultural Academy[1]), Lysenko revealed that Stalin himself had endorsed his position in biology; Kedrov was forced to admit error in having opposed Lysenko and *Voprosy filosofii*'s anti-Lysenko line was reversed. The elder Zhdanov went into retirement and died shortly thereafter.

The third debate—in economic science—was not a clear factor in Zhdanov's fall, but the course of this debate also reflected the swing toward conservatism in late 1947. Zhdanov had not intervened in economics, which was the preserve of his ally N. A. Voznesenskiy. Still, Varga's view that Western capitalist countries had adopted some socialist elements and were becoming more like the Soviet Union suggested that the West was not so hostile and that the Soviet Union could coexist and collaborate with capitalist countries. This view accorded with the belief, apparently held by Zhdanov and even Stalin during 1946, that peaceful coexistence would continue. Perhaps reassured by this implicit support at the very highest levels, the leading economists defended Varga, and conservative criticism of his economic views made little headway until late 1947.

As the cold war developed later in 1947, Varga's views were no longer compatible with Stalin's position, and Zhdanov's ambitious ally Voznesenskiy took up the attack on Varga for presenting such a pragmatic, overly sympathetic view of capitalism. During 1948, the attacks on Varga by Voznesenskiy and others became linked with the attacks on Zhdanov-sponsored liberals and moderates in philosophy and biology, and even as Zhdanov fell, Voznesenskiy's influence continued to grow. The anti-Varga campaign, however, did not have such severe consequences as the crackdown in philosophy and biology—probably because it had been spearheaded by Zhdanov's ally Voznesenskiy and slackened when Voznesenskiy himself was purged in early 1949. Despite Varga's apparent opposition to the cold war, the fact that Zhdanov's foes made relatively little attempt to punish Varga after Zhdanov's death suggests that the economist had no close tie to Zhdanov.

The Philosophy Debate

The great philosophy debate of 1947 was forced by Stalin's personal intervention and provided an opening through which hard-pressed dogmatists could strike back at Zhdanov's protégés. The

---

[1] *Vsesoyuznaya akademiya selskokhozyaystvennykh nauk imeni Lenina.*

philosophers were roughly divided into an older, dogmatic group (M. B. Mitin, aged 46, P. F. Yudin, 48, and A. A. Maksimov, 56), which had lost most of its power in a 1942–44 conflict, and a younger, more flexible group (G. F. Aleksandrov, aged 39, P. N. Fedoseyev, 39, M. T. Iovchuk, 39, B. M. Kedrov, 44, G. S. Vasetskiy, 43), which had taken over virtually all the leading posts in philosophy.[2] The dogmatists eventually won Stalin's favor and during 1949 totally displaced the Zhdanov group.

The 1944 conflict had arisen over the publication of the third volume of a *Istoriya filosofii* (History of philosophy), edited by Yudin and Mitin. Errors discovered in this volume were apparently used as a pretext to discredit Yudin and Mitin, remove them from their leading posts, and abolish their ideological journal, *Pod znamenem marksizma.* As head of OGIZ and director of the Institute of Philosophy, Yudin had rushed through the third volume in hopes of winning a Stalin Prize.[3] Mitin, as chief editor of *Pod znamenem marksizma* and deputy director of the Institute of Philosophy, had protected the new volume from criticism.[4] Initially, they were successful. The March 1943 *Vestnik* of the Academy of Sciences announced that the three-volume *Istoriya filosofii* (published 1940–42) had been awarded the Stalin Prize for 1942. An article in the March 1943 *Pod znamenem marksizma* boasted that this was the first time a work of philosophy had won a Stalin Prize.

Praise for the third volume stopped abruptly, however, when an editorial article in an April issue of *Bolshevik* (no. 7–8, 1944, signed to press 15 May 1944), entitled "On shortcomings and errors in describing the history of German philosophy of the end of the 18th and beginning of the 19th centuries," attacked the third volume for

---

[2] The idea that the philosophers participating in the June debate were divided according to age has been raised by both Zhdanov and Kedrov. Zhdanov's speech in the June debate noted that "representatives of the older generation of philosophers" were reproaching "some young ones" (*Voprosy filosofii*, no. 1, 1947, p. 269). Kedrov, in the December 1973 *Voprosy filosofii*, noted that he had proposed the creation of *Voprosy filosofii* during the June 1947 debate and had been supported by "the young philosophers" (p. 156).

[3] According to Fedoseyev's speech in the June 1947 philosophy debate (*Voprosy filosofii*, no. 1, 1947, p. 456).

[4] In his speech in the June 1947 debate, Z. A. Kamenskiy noted that he had sent a review of the third volume to the Institute of Philosophy and the editors of *Pod znamenem marksizma,* and that the leaders of these organs—the same ones who had edited the third volume—had deleted all criticism from the review (*Voprosy filosofii*, no. 1, 1947, p. 377).

failing to expose the fact that Hegel had been a blatant German racist and nationalist. The article accused the authors of overlooking Hegel's statement that wars played a positive role in unifying and strengthening countries, as well as his notion that the German people were a chosen people called upon to rule other peoples, apparently including the Slavic peoples, of whom Hegel had held a low opinion. The *Bolshevik* article announced that the Stalin Prize Committee had reconsidered its earlier award and now had decided that the prize applied only to the first two volumes. This article was reprinted in an issue of *Pod znamenem marksizma* signed to press on 8 June 1944 (no. 4–5, 1944).[5] According to speeches delivered by Kedrov[6] and Iovchuk[7] during the 1947 debate, the CC had condemned the volume for errors and had named a new editorial board, including Fedoseyev, Kruzhkov, and Svetlov, to write a revised version.

Although Aleksandrov had been the formal head of the disgraced volume's editorial board, the 1947 debate attributed the major role to his fellow editors Yudin and Mitin, thus providing a clear pretext for attacking their journal as well.[8] Yudin and Mitin had long led the editorial board of *Pod znamenem marksizma*,[9] but the same issue that reprinted the *Bolshevik* attack listed Iovchuk (who had not previously been on the editorial board) as chief editor. After this issue, the journal ceased publication and thereafter was treated as having been a failure, a point repeatedly cited to discredit Yudin and Mitin during the 1947 philosophy debate.[10]

The purge of Yudin and Mitin at *Pod znamenem marksizma* was accompanied by their ouster from the leadership of the Institute of Philosophy, where Yudin was director and Mitin deputy director. According to Aleksandrov (*Bolshevik*, no. 14, July 1945, p. 22), a 1944 CC decree had assailed not only the errors in the third volume but also the work of the Institute of Philosophy; soon thereafter "new people came into the leadership" of the institute. Yudin was

[5] Mitin, commenting on the *Bolshevik* editorial article in a June issue of *Bolshevik* (no. 12, 1944), followed its line and assailed Hegel as a German nationalist and apologist of war.

[6] *Voprosy filosofii*, no. 1, 1947, p. 38.

[7] Ibid., p. 221.

[8] Aleksandrov himself harshly attacked the third volume in an article in a July issue of *Bolshevik* (no. 14, 1945, pp. 21–22).

[9] Their ally Maksimov also had long been a member but left the journal in 1942.

[10] In the June debate Zhdanov spoke of the "sad experience" with the journal *Pod znamenem marksizma* (*Voprosy filosofii*, no. 1, 1947, p. 267), and P. Ye. Vyshinskiy spoke of the journal's "errors" (ibid., p. 231).

reduced to head of the department of Marxism-Leninism at Moscow State University, and Mitin was demoted to leader of the department of philosophy of the Higher Party School, losing his post as director of the Marx-Engels-Lenin Institute, as well as the jobs of chief editor of *Pod znamenem marksizma* and deputy director of the Institute of Philosophy.[11] Yudin was further demoted by Zhdanov in 1946. In the midst of Zhdanov's campaign, a CC decree (5 October 1946) assailed the work of OG IZ and fired its head Yudin for irresponsible leadership.[12]

Yudin and Mitin were succeeded at the Institute of Philosophy by V. I. Svetlov and M. P. Baskin, who in turn were replaced by G. S. Vasetskiy and B. M. Kedrov.[13] The new leaders used their positions as Yudin and Mitin had. When Agitprop chief Aleksandrov wrote his 1946 *Istoriya zapadnoyevropeyskoy filosofii* (History of Western European philosophy), it was promoted with the active assistance of Kedrov and critics of the book were suppressed.[14] The book was widely praised and quickly awarded a Stalin Prize.[15]

Sometime in late 1946 or possibly in early 1947, however, Stalin suddenly criticized the book for shortcomings. The nature of Stalin's objections is unclear since the criticism was never made public. Perhaps the most important theme in the subsequent criticism was Zhdanov's statement that by not including Russian philosophy in this book about Western philosophy, Aleksandrov, intentionally or not, had belittled the role of Russian philosophy—a criticism Alek-

---

[11] See his biography in the *Filosofskaya entsiklopediya*, vol. 3.

[12] The decree is published in the 1971 edition of *KPSS v rezolyutsiyakh i resheniyakh syezdov, konferentsiy i plenumov TsK*, vol. 6, p. 190, and in the 30 October 1946 *Kultura i zhizn*. OG IZ and Yudin had also been attacked for poor work in the 10 July 1946 *Kultura i zhizn*. *Kultura i zhizn* on 20 November 1946 announced that Yudin had just been appointed chief editor of the Academy of Sciences journal *Sovetskaya kniga*. He began a comeback in 1947, however. According to his biography in the second edition of the *Bolshaya sovetskaya entsiklopediya,* he became chief editor of the main trade-union organ, *Trud,* in 1947.

[13] See Kamenskiy's speech in *Voprosy filosofii,* no. 1, 1947, pp. 377, 379. Vasetskiy's appointment was announced in the *Vestnik akademii nauk SSSR,* July 1946.

[14] In the June 1947 debate, the conservative A. A. Maksimov declared that Institute of Philosophy first deputy director S. L. Rubinshteyn had opposed submitting Aleksandrov's book for a Stalin Prize and therefore the matter was taken out of his hands and entrusted to Kedrov. Rubinshteyn soon quit the institute (*Voprosy filosofii,* no. 1, 1947, pp. 190–91).

[15] In the June debate, N. M. Miroshkina noted the praise for the book and observed that the book was distributed all over the country and used as a textbook (ibid., p. 147).

sandrov acknowledged as correct.[16] Although it was Zhdanov who raised this theme, it is not clear whether it reflected his own views or was primarily influenced by Stalin's sentiments and the new nationalist tide—much like Zhdanov's September 1947 "two camps" speech, which represented a marked departure from his 1946 position.[17]

In connection with Stalin's criticism, the CC ordered the Institute of Philosophy to hold a discussion of Aleksandrov's book.[18] At the initial discussion, held in the Academy of Sciences in January 1947, Zhdanov's clique tried to cover up for Aleksandrov and keep the criticism private.[19] Zhdanov, in opening the second discussion of the book (16 June 1947), declared that the CC considered the results of the January debate "unsatisfactory," because too few persons had had an opportunity to speak and because only edited excerpts of the speeches, not verbatim accounts, had been published.[20] Yudin, in his speech at the June 1947 meeting, went further, claiming that the organizers of the January debate—P. N. Fedoseyev, V. S. Kruzhkov, and G. S. Vasetskiy—had tried to limit criticism of Aleksandrov's book.[21] Indeed, since all three men were Aleksandrov's subordinates (Fedoseyev as deputy chief of Agitprop, Kruzhkov as director of the Marx-Engels-Lenin Institute, and Vasetskiy as director of the Institute of Philosophy) and had aided in the publication of the book, such an action on their part would have been understandable.

The second discussion lasted from 16 June to 25 June, with Zhdanov presiding. This time full versions of all the speeches were published for everyone to read—501 pages in the first issue of the new journal *Voprosy filosofii* (no. 1, 1947). The second debate gave

---

[16] Ibid., p. 260, 294.

[17] Despite Zhdanov's statement, however, Kedrov argued that Aleksandrov's mistake was that he had precisely tried to exaggerate the role of Russian philosophy and underrate international class analysis in favor of national features (ibid., p. 48). Zhdanov did not appear upset by Kedrov's statement: speaking after Kedrov and commenting on Kedrov's ideas, he did not object to Kedrov's views on favoring the international over the national, and, in fact, right after the June meeting he followed Kedrov's advice in creating the new journal *Voprosy filosofii* and made Kedrov its chief editor.

[18] See the speeches of Zhdanov (ibid., p. 267), M. D. Kammari (ibid., p. 16), G. M. Gak (ibid., p. 25), and Yudin (ibid., p. 280) in the June 1947 discussion.

[19] No account of the discussion appeared in the academy's *Vestnik* or in the journal of its division of history and philosophy (*Izvestiya akademii nauk: seriya istorii i filosofii*).

[20] *Voprosy filosofii*, no. 1, 1947, p.5.

[21] Ibid., pp. 279–90.

the Yudin-Mitin group its chance to criticize the errors in the new history of philosophy and use them to turn the tables on their rivals. Whatever Stalin's criticism of the book, the discussion soon turned into a debate between the "older" and "younger" philosophers. Kedrov, while criticizing the Aleksandrov book, brought up the errors of the 1944 Yudin volume, claiming that in attacking Aleksandrov "some comrades" were dragging us back to these discredited volumes that had also been exposed as erroneous by Stalin.[22] Mitin responded by declaring that some people still did not really want to criticize Aleksandrov and he assailed Kedrov's early 1947 book, *O kolichestvennykh i kachestvennykh izmeneniyakh v prirode* (On qualitative and quantitative changes in nature), for mistakes similar to Aleksandrov's.[23] Maksimov then attacked Kedrov for having blocked criticism of Aleksandrov's book by Rubinshteyn, first deputy director of the Institute of Philosophy.[24] Iovchuk then criticized Mitin and recalled the CC condemnation of the 1944 history of philosophy.[25] P. Ye. Vyshinskiy, discounting Rubinshteyn's opposition to the Aleksandrov book and recalling the errors of *Pod znamenem marksizma,* asked Mitin why he, as a member of the Stalin Prize Committee, had not objected to awarding a prize to the book. Noting Mitin's criticism of Kedrov's book, Vyshinskiy criticized Mitin for calling Kedrov a "menshevizing idealist."[26] Z. A. Kamenskiy characterized the role of Mitin and Yudin as simply "defending orthodox Marxism against distortions," without making any original scholarly contributions.[27] I. A. Kryvelev then assailed Mitin for producing nothing himself, even while attacking Kedrov, who at least "is a working and creative person."[28] Yudin in turn accused Fedoseyev, Kruzhkov, and Vasetskiy of having tried to protect Aleksandrov's book from criticism.[29] Fedoseyev countered that the only work Yudin had issued was the ill-fated 1944 third volume, which he had rushed through in order to win a Stalin Prize.[30]

[22] Ibid., p. 38.
[23] Ibid., pp. 122–26.
[24] Ibid., pp. 190–91.
[25] Ibid., p. 221.
[26] Ibid., pp. 227–31.
[27] Ibid., p. 376.
[28] Ibid., p. 393.
[29] Ibid., pp. 279–81.
[30] Ibid., p. 456. The speeches of Kamenskiy, Kryvelev, and Fedoseyev were not actually delivered because of the cutoff of discussion on 25 June; however, they were published with the other speeches in *Voprosy filosofii,* no. 1, 1947.

When Aleksandrov finally spoke, he admitted errors in his book but used his speech to assail Mitin and Yudin bitterly, noting in particular Yudin's "many years of theoretical fruitlessness."[31]

The dogmatists broadened the attack to Agitprop, claiming that a clique was monopolizing all leading posts. This charge was first raised by D. I. Chesnokov, the conservative Sverdlovsk philosopher, who, instead of just attacking the Institute of Philosophy or other philosophers, placed the blame directly on Agitprop and specified its leaders: Aleksandrov, Fedoseyev, and Iovchuk.[32] Chesnokov was seconded by Yudin, who complained that Fedoseyev and his group were monopolizing all top posts in philosophy and that people were chosen "on the group principle." He noted that Fedoseyev was not only an Agitprop leader but also head of the philosophy department in the Academy of Social Sciences and editor of *Bolshevik*. Yudin declared that if the philosophy department of the Academy of Social Sciences (headed by Fedoseyev) is working poorly, one has to complain to the CC Propaganda Administration (which Fedoseyev helps lead), and if you have a conflict with *Bolshevik* chief editor Fedoseyev, you have to complain to the Propaganda Administration (again Fedoseyev).[33]

Toward the end of these discussions, on 24 June, Zhdanov delivered a speech in which he harshly criticized Aleksandrov for personal shortcomings as well as errors in his book. He admitted that a small group was monopolizing philosophy and noted that the older philosophers were rebuking the younger for lack of militancy.[34]

This monopoly soon ended. After the June 1947 debate, Aleksandrov was removed as chief of Agitprop and as chief editor of *Kultura i zhizn*.[35] He replaced Vasetskiy as director of the Institute of Philosophy. Fedoseyev also left Agitprop at this time, although he

---

[31] Ibid., p. 298.

[32] Ibid., p. 245. Though relatively young (thirty-seven), Chesnokov clearly was part of the Mitin-Yudin group and prospered after Zhdanov's death. He replaced the purged Kedrov as *Voprosy filosofii* chief editor in early 1949 and even became a CC Presidium member in 1952. In the June 1947 debate he spoke as representative of Sverdlovsk, according to *Voprosy filosofii*. At that time, according to his biography in the *Filosofskaya entsiklopediya* (vol. 5, 1970), he was a secretary of the Sverdlovsk city party committee.

[33] *Voprosy filosofii,* no. 1, 1947, p. 286.

[34] Ibid., p. 269.

[35] He was last identified as chief of Agitprop in the 30 June 1947 *Kultura i zhizn.* (Also see his biography in *Filosofskaya entsiklopediya,* vol. 1.) He last signed as chief editor of *Kultura i zhizn* on 10 September 1947.

retained two other posts: *Bolshevik* chief editor and chief of the department of dialectical and historical materialism at the Academy of Social Sciences.[36] Iovchuk had left Agitprop when transferred to Belorussia as propaganda secretary in March 1947.

A new group of leaders soon took over from the Zhdanovites at Agitprop. M. A. Suslov was identified as chief of the Propaganda Administration in the 23 November 1947 *Moskovskiy bolshevik* and D. T. Shepilov as deputy chief in the 21 November 1947 *Izvestiya*. A second new deputy chief, L. F. Ilichev, took office early in 1948.[37] Succeeding Aleksandrov as chief editor of Agitprop's organ *Kultura i zhizn* was deputy editor P. A. Satyukov (who, however, appeared to be a Zhdanov protégé from Gorkiy[38]). One of the Agitprop deputy chiefs was not a Zhdanov man and was not replaced during this change: V. G. Grigoryan, a protégé of Beriya.[39]

Although the moderate clique was thus on the defensive during the June 1947 debate and suffered serious setbacks immediately afterward, it did win one very notable victory. During the debate, Kedrov appealed to Zhdanov personally to help create a journal especially for philosophers.[40] His request was repeated by P. Ye. Vyshinskiy, as well as V. I. Svetlov and G. S. Vasetskiy, respectively the former and current directors of the Institute of Philosophy. Zhdanov was reluc-

---

[36] His last press identification as Agitprop deputy chief was in the 13 January 1947 *Pravda*, although judging by statements in the debate he was still there in June 1947. *The Bolshaya sovetskaya entsiklopediya* yearbook for 1971 lists him in the CC apparat from 1941 to 1947. He was fired as *Bolshevik* chief editor by a 13 July 1949 CC decree. Fedoseyev headed the Academy of Social Sciences' dialectical and historical materialism department, rather than the philosophy department, as Yudin claimed in the above statement.

[37] Ilichev, still identified as chief editor of *Izvestiya* in the 4 December 1947 *Vechernyaya Moskva*, was first identified as Agitprop deputy chief in the 24 March 1948 *Pravda*.

[38] Satyukov had started his career working on the Gorkiy *oblast* paper in 1930, while Zhdanov was *oblast* first secretary, and he rose to chief editor of the paper. In the 1940s, while still in his early thirties, he became head of a CC section, apparently in Agitprop, and when *Kultura i zhizn* was established in 1946, he became its deputy editor (see his biographies in the *Bolshaya sovetskaya entsiklopediya* yearbook for 1962 and the 1962 *Deputaty verkhovnogo soveta SSSR*). The 1962 yearbook lists him as deputy editor, then editor of *Kultura i zhizn*, 1946–49.

[39] For many years, Grigoryan was chief editor of *Zarya vostoka*, the leading paper in Beriya's bailiwick of Georgia, but he left this post on 27 April 1946, shortly after Beriya's March 1946 promotion to the Politburo, and by 10 July 1946 was writing in *Kultura i zhizn*, which suggests that he held a post in Agitprop. On 11 January 1947, 11 February 1947, and 30 April 1947, *Kultura i zhizn* identified him as deputy head of Agitprop.

[40] *Voprosy filosofii*, no. 1, 1947, p. 52.

tant at first, citing the "sad experience" of *Pod znamenem marksizma*, [41] and Mitin and Yudin ignored the proposal in their speeches. It was the young philosophers who supported the proposal to create *Voprosy filosofii* at this meeting, according to Kedrov in a December 1973 interview in *Voprosy filosofii*. [42] Despite initial hesitancy, Zhdanov soon changed his mind, however, and the CC quickly created such a journal. Moreover, Kedrov himself was named chief editor, and P. Ye. Vyshinskiy and I. A. Kryvelev, both of whom had defended Kedrov against Mitin, became members of its editorial board. As Kedrov said in the December 1973 interview, "we were in seventh heaven when we found out that the CC had sanctioned the creation of the new journal." But, he added, this situation "unfortunately did not last very long—just to the beginning of 1949." [43]

The Biology Debate

At the same time as the philosophy debate sparked by Aleksandrov's book, an even more violent controversy was occurring in biology, and once again, it was charged that certain figures had unpatriotically underrated the achievements of Russian science. Trofim Denisovich Lysenko, who since the mid-1930s had enjoyed Stalin's favor and was striving to force his pseudoscientific theories on Soviet biology, argued that Michurinism (that is, Lysenko's biological theories) was a great achievement of native Russian science far superior to the theories of genetics propounded by Western scientists. In Lysenko's view, then, any Soviet scientist who challenged Michurinism was antipatriotic and servile to the West. Lysenko labeled his foes "Morganists-Weismannists," that is, followers of the American Thomas Hunt Morgan and the German August Weismann, pioneers in the field of genetics.

In his accounts of the rise of Lysenko, Soviet biologist Zhores A. Medvedev, who was very well informed on developments in Soviet biology in the postwar period, has written that Lysenko was struggling to promote his theories during 1947, but, despite support

[41] Ibid., p. 267.

[42] Ibid., no. 12, 1973, p. 155.

[43] Ibid., p. 156. After Zhdanov's death in mid-1948, Kedrov immediately came under fire and soon was removed. Chesnokov became chief editor and Mitin and Maksimov also joined *Voprosy filosofii*'s editorial board. Under their leadership, the journal assailed Kedrov for defending "cosmopolitan" views.

from Stalin and other Soviet leaders, failed to win wide support among scientists.[44] Medvedev notes, for example, that the scheduled October 1947 election of new members of VASKhNIL was canceled when it became clear that a majority of members opposed Lysenko (then president of VASKhNIL) and would not vote for his supporters.[45]

Lysenko's campaign, ignored by scientific publications such as the *Vestnik* of the Academy of Sciences, developed instead in *Literaturnaya gazeta,* the organ of the Writers' Union. Soon after the June 1947 philosophy debate, the dogmatic Mitin was added to *Literaturnaya gazeta*'s editorial board[46] where, according to an editorial note in *Voprosy filosofii,* he was the editor in charge of science.[47] Mitin soon turned *Literaturnaya gazeta* into a vehicle for Lysenko's views, and in doing so he clearly had the backing of V. V. Yermilov, the chief editor. Even though Yermilov had been appointed to this post in November 1946, at the height of Zhdanov's campaign, he was personally attacking Kedrov and other foes of Lysenko by 1948 (see below).

On 18 October 1947, *Literaturnaya gazeta* carried an interview with Lysenko, in which he defended his theory of the impossibility of intravarietal competition and promoted the cluster or nest method of planting, which had supposedly been invented by I. D. Kolesnik.[48] He argued that this theory was an original contribution by Russian science and that since Western biologists could not accept it for ideological reasons, they were attacking it. He linked his Russian scientific foes with foreign scientists, singling out the prominent

---

[44] In 1962 Medvedev wrote accounts of Lysenko's rise and purge of biologists which circulated among Soviet scientists and eventually got abroad and were published in the February and May 1969 issues of the émigré journal *Grani* (nos. 70 and 71). I. M. Lerner translated parts of this manuscript, plus additions written in 1963–67, and published them as Zh. A. Medvedev, *The Rise and Fall of T. D. Lysenko* (New York, 1969).

[45] Medvedev, *The Rise and Fall of T. D. Lysenko,* pp. 115–16.

[46] He was first listed on the editorial board in the 30 August 1947 issue of the newspaper.

[47] *Voprosy filosofii,* no. 1, 1948, p. 225.

[48] Lysenko believed that plants did not compete for existence with other plants of the same variety or species but only with plants of a different variety. Hence, one could plant a number of seeds of one type of plant close together (the nest or cluster method) and some would voluntarily die out to allow the others to flourish. After Lysenko's triumph, a huge program of planting forest belts (Stalin's plan for the transformation of nature) was announced in late 1948, using the nest method. However, the trees did not naturally space themselves, as Lysenko's theory dictated, but choked each other off and died.

biologist P. M. Zhukovskiy (director of the Plant Growing Institute in Zhdanov's Leningrad) for ridiculing his theory that there was no competition within a given variety, and Professor A. A. Sokolov for ridiculing Kolesnik and claiming that his nest method discovery had been tried in the United States as many as ten years earlier.

Subsequently, as the *Literaturnaya gazeta* of 29 November 1947 explained, the Moscow State University biology faculty held a large conference to discuss Lysenko's views. Their evaluation was clearly negative and professors I. Shmalgauzen, A. Formozov, D. Sabinin, and S. Yudintsev sent an article to *Literaturnaya gazeta* objecting to Lysenko's views. On 29 November 1947, the paper published the article, but simultaneously ran a pro-Lysenko piece by scientists A. Avakyan, D. Dolgushin, and others. On 10 December 1947, *Literaturnaya gazeta* printed pro-Lysenko articles by V. Stoletov and N. Turbin and an anti-Lysenko article by B. Zavadovskiy. The latter complained that "there exists the incorrect idea that anyone who criticizes Lysenko thereby joins in a common front with formal geneticists and other Morganists, one-sidedly denying that T. D. Lysenko has any kind of scientific merits at all." The debate concluded with a long article by Mitin (27 December 1947), in which he defended Lysenko, assailed Zavadovskiy and Shmalgauzen, and also attacked biologist A. Zhebrak for publishing an article in the American journal *Science,* thus joining with "reactionary" American scientists in their attempt to discredit Michurinism.

Only four days later—on 31 December 1947—*Kultura i zhizn* entered the debate, carrying an article by A. A. Maksimov which attacked Kedrov's 1947 book *Engels i yestestvoznaniye* (Engels and natural science) for serious errors. This attack apparently was facilitated by the shift in control of Agitprop. Suslov had become chief of Agitprop and P. A. Satyukov had just become chief editor of the Agitprop organ *Kultura i zhizn.*

Kedrov's book also became the subject of sharp criticism at a January 1948 debate at the Institute of Philosophy, according to an article by D. Ivanenko in *Bolshevik* (no. 8, 1948). In this issue of *Bolshevik,* signed to press on 30 April 1948, Ivanenko claimed that Kedrov's book on Engels neglected the "achievements of Russian science" and "the question of the priority of Soviet scientists, whose works are belittled by some bourgeois scholars."

The 8, 12, and 16 January 1948 debate at the Institute of Philosophy was described in detail in the March 1948 *Vestnik* of the

Academy of Sciences (signed to press 20 March 1948). The account reported that Maksimov had repeated his *Kultura i zhizn* criticisms of Kedrov's book and harshly assailed it for "conclusions hostile to Marxism." The other speakers only mildly criticized the book and many (B. G. Kuznetsov, V. G. Fridman, M. A. Markov, G. S. Vasetskiy) assailed Maksimov for his attacks on it. Aleksandrov, who presided, criticized the book for overemphasizing the international character of science but also attacked some speakers (clearly meaning Maksimov) for attributing nonexistent errors to Kedrov. Kedrov lashed back at Maksimov. The account clearly indicated that Kedrov was not in serious trouble and that Maksimov was fighting a lonely, uphill struggle against Kedrov. As the March *Vestnik* editorial summed up the debate, most speakers had found errors in the book but also had stressed its "great positive significance" and had rejected the accusations of some of Kedrov's foes.

The March *Vestnik* editorial also praised the debate over intravarietal struggle then being conducted in *Literaturnaya gazeta*, but indicated that it had not yet been resolved. The journal carried the decision of the bureau of the Academy of Sciences' division of biological sciences after a 11 December 1947 debate on the subject, in which Lysenko, Shmalgauzen, and others had participated. The bureau decision leaned slightly to Lysenko's foes, declaring that struggle between plants of the same variety does in fact occur (which Lysenko denied) and rebuking those who accused others of "Malthusianism" or "Lamarckism."[49] Nevertheless, the bureau tried to avoid a clear decision, simply calling for further study.

Meanwhile, Kedrov was using *Voprosy filosofii* to fight Lysenko and the conservative philosophers. The new journal's first issue was taken up entirely with the speeches that had been delivered at the June 1947 debate, but the subsequent issues boldly ignored Lysenko and carried provocative, unorthodox articles on science and attacks on the conservatives. The second issue for 1947 (signed to press only on 2 February 1948) opened with an editorial criticizing Mitin for lack of serious output, and included articles by Kedrov, I. A. Kryvelev, M. A. Markov, Z. A. Kamenskiy, and I. I. Shmalgauzen — all of which were later attacked sharply by the conservatives.

Literaturnaya gazeta quickly pounced on *Voprosy filosofii*, carrying an article by Maksimov (10 April 1948) which assailed Mar-

---

[49] "Malthusianism" was an epithet used by Lysenko, whereas his foes accused him of "Lamarckism."

kov's article on physics. Maksimov declared that Markov's whole article was "an apology" for the philosophical views of the Danish physicist Niels Bohr and was trying to make Soviet readers swallow Bohr's reactionary Western philosophical views along with his views on physics. *Literaturnaya gazeta*'s editors added an introductory note to Maksimov's article questioning whether *Voprosy filosofii* was accurately reflecting the lessons of the June 1947 philosophy debate. The second issue of *Voprosy filosofii* also stirred a discussion in the Writers' Union. According to a 19 May 1948 *Literaturnaya gazeta* report, Kamenskiy's article was criticized and during the debate, Kedrov refused either to defend or criticize it.

Kedrov attacked Maksimov's *Literaturnaya gazeta* article during a 19 April meeting of the Academy of Sciences' editorial council, and accused Maksimov of arguing that Leninism in the philosophy of natural science represented a continuation of *Russian* science rather than a further development of Marxism.[50] In addition, Kedrov used the next issue of *Voprosy filosofii* (no. 1, 1948, signed to press 29 June 1948) to attack Maksimov and *Literaturnaya gazeta* strongly, while defending Markov. In a signed article Kedrov attacked Maksimov for himself "bowing before things foreign" and for having praised Aleksandrov's book in 1946. In addition, a note from the editors explained that Maksimov had first sent his article attacking Markov to *Voprosy filosofii* and that *Voprosy filosofii*, while disagreeing with it, had agreed to print it along with a refutation. The note accused Maksimov of deceit in having the article printed in *Literaturnaya gazeta* when the article had already been set for *Voprosy filosofii* (no. 1, 1948). Furthermore, it complained that *Literaturnaya gazeta*'s editorial introduction to Maksimov's article had accused *Voprosy filosofii* of propagating revision of Marxism and claimed that the attack on the Markov article was "in fact" directed at *Voprosy filosofii*'s editors.

Literaturnaya gazeta immediately struck back. An unsigned article about "serious errors" of *Voprosy filosofii* in the 17 July 1948 issue accused the editors of ignoring the ideological struggle and "not printing even one serious article that examined the problem of Soviet patriotism and exposed manifestations of bourgeois cosmopolitanism and bowing before the culture of the West." The article placed the blame on Kedrov, who in his 1947 book on Engels

---

[50] Reported in the June 1948 *Vestnik* of the Academy of Sciences.

had denied "the importance of the question of the priority of the fatherland's science," thereby offending the Soviet people, "who are rightly proud of Russian primacy in the invention of the steam engine, electric lamp, radio, and in creating the first airplane—the huge contribution of our scientists to the treasure house of world culture." Kedrov had cosmopolitan views, it argued, because he had a "nihilistic attitude to questions of national culture" and a "scornful attitude" toward early Russian philosophy. *Literaturnaya gazeta* also assailed as an "antipatriotic act" Professor Zhebrak's publication of an anti-Lysenko article in the American journal *Science,* and denounced the "antiscientific and conservative views of representatives of formal genetics," such as I. Shmalgauzen, A. Formozov, S. Yudintsev, and D. Sabinin. The article declared that *Voprosy filosofii* should be defending "Soviet agrobiological science" (that is, Lysenkoism) and attacking those scientists who bow and scrape before Western genetics. *Literaturnaya gazeta* chief editor Yermilov followed this blast with a long signed article (*Literaturnaya gazeta,* 29 August 1948), in which he assailed Kedrov's cosmopolitan views.

The debate between *Literaturnaya gazeta* and *Voprosy filosofii* was resolved with Zhdanov's loss of power. The October 1948 issue of the *Vestnik* of the Academy of Sciences reported an Institute of Philosophy discussion of the August VASKhNIL session that had endorsed Lysenko's views. The meeting, held on 13 August, was run by G. F. Aleksandrov and D. I. Chesnokov, now director and deputy director of the institute, respectively. Mitin assailed *Voprosy filosofii* for carrying an article by Shmalgauzen, the leader of the "Weismannists-Morganists," and ignoring the ideological war in biology. Kedrov confessed to error in not combating the geneticists and admitted that it had been a mistake to print Shmalgauzen's article. On 27 August 1948, *Pravda* reported a 26 August decree of the Academy of Sciences Presidium which criticized the Institute of Philosophy for its failure to support Lysenko.

The next issue of *Voprosy filosofii* (no. 2, 1948) appeared only after Zhdanov's death and the August 1948 VASKhNIL session that endorsed Lysenko's views. Signed to press on 27 October 1948, this issue marked a complete surrender to *Literaturnaya gazeta* and the conservatives. Although Kedrov remained the nominal chief editor, the lead editorial ("Against the bourgeois ideology of cosmopolitanism") attacked Kedrov's book on Engels for undervaluing Russian science. The editorial noted that the error of Aleksandrov's book was to belittle Russian philosophy, and it accused the Kamenskiy article on eighteenth- and nineteenth-century Russian philosophy (*Voprosy filosofii*, no. 2, 1947) of cosmopolitanism in taking a similar line. The same editorial assailed Shmalgauzen's *Voprosy filosofii* article, claiming that it followed the "metaphysical and idealistic theories of foreign bourgeois biologists," instead of relying on the successes of "Soviet biology headed by Academician T. D. Lysenko." Such cosmopolitan biologists, the editorial complained, bow before the big names of foreign science and "fail to see that our Russian, Soviet science achieves victory after victory" in biology. The same issue contained articles lauding Lysenkoism by I. Ye. Glushchenko and V. N. Stoletov (both Lysenkoists), and even included a greeting to Lysenko on his fiftieth birthday.

The next issue (no. 3, 1948, signed to press 1 June 1949) attacked the Markov, Kamenskiy, Kryvelev, and Shmalgauzen articles, declared Kedrov's early 1948 article (*Voprosy filosofii*, no. 2, 1947) deeply mistaken, and assailed Kedrov for defending "cosmopolitan viewpoints." Kedrov was removed as chief editor and Kryvelev deposed as responsible secretary of the editorial board.

The Economics Debate

In the third debate, that on economic science, conservatives also challenged the scholarly establishment, specifically asserting that it was too soft in its attitude toward the West. The debate focused on a 1946 book by Ye. S. Varga, *Izmeneniya v ekonomike kapitalizma v itoge vtoroy mirovoy voyny* (Changes in the economy of capitalism as a result of the Second World War). Varga, director of the Institute of World Economics and World Politics and member of the Presidium of the Academy of Sciences, was perhaps the Soviet Union's top economist. His book argued that during the war capitalist governments had acquired great power over their economies and that the elements of planning, introduced to help focus economies on the war effort, had been retained after the war. Such an analysis implied that planning enabled capitalist states to deal more effectively with their economic crises and thus become stronger. At the same time, moreover, Varga's view suggested that capitalist states were becoming more like socialist states, a circumstance that perhaps rendered them less objectionable from the Soviet point of view.

These implications were highly distasteful to Soviet conservatives, who looked for extreme, perhaps fatal, crises in the West after the war. They argued that monopolies were dominant in the West and that governments, as mere instruments of these monopolies, were not powerful enough to control the disruptive economic forces—a situation that would lead to economic disaster. The analyses of Varga and his colleagues, however, presented Western states as economically sound after the war. This view and their favorable attitude toward Western economic reforms led to charges that they really favored "reformism" over revolution.

The Varga debate began in May 1947, when the political economy sector of the Institute of Economics and Moscow State University's political economy department held a joint meeting to discuss Varga's book. At this point, politics had not entered the subject and the debate was scholarly and two-sided, with varied opinions freely discussed.[51] During the discussion, some conservatives challenged Varga's thesis that Western governments had developed decisive control of their economies during the war and were now able to control economic conditions sufficiently to avoid economic disasters. Making the most politically charged comments, A. I. Kats of the Institute of Economics accused the book of downplaying the crisis of capitalism and instead portraying capitalism as on the upswing after the war. V. Ye. Motylev of Moscow State University also claimed that Varga's book presented a distorted picture of prospects of capitalist countries. In his opening speech, however, Varga first reasserted the arguments he had made in his book and then stated flatly that elements of planning had developed in capitalist states during World War II, that they continued to exist after the war, and that he would not now write a book in any way different from the one he had finished in 1945. He declared that it was simply wrong to assert that there is no planning in Western economies.

Throughout the May debate, most speakers sided with Varga, making certain academic criticisms of his book, but praising its innovative character. K. V. Ostrovityanov, deputy academic secretary of the Academy of Sciences' division of economics and law, set the tone with his opening remarks, which lauded the book as bold

[51] The stenographic report of the speeches during the May debate was printed in a supplement to the November 1947 issue of *Mirovoye khozyaystvo i mirovaya politika*, the organ of Varga's institute.

and thought-provoking despite some incorrect propositions and called for "comradely" discussion and constructive criticism. Other speakers not only praised Varga's book but also assailed the conservatives' criticism and defended the need to explore new ideas. M. I. Rubinshteyn (a member of the editorial board of Varga's journal) accused Motylev of adopting a "prosecutor's" tone toward the book and he complained that Kats, who attacked Varga for failing to talk about the "crisis of capitalism," was so preoccupied with the concept of "crisis of capitalism" that he never contributed any new ideas. V. A. Maslennikov, deputy director of the Institute of Economics and also a member of the editorial board of Varga's journal, declared that Kats sought to discredit Varga because he had contradicted Kats's thesis that capitalism was now entering a sharp downturn. M. A. Arzhanov asserted that the reason some did not like Varga's book was that it presented a realistic view of the West instead of claiming that the collapse of capitalism was imminent. As Ostrovityanov put it in his concluding remarks, the "overwhelming majority" of speakers considered the book a "major scientific and theoretical" work, and hardly anyone tried to write it off as unimportant.[52] In his concluding statement, Varga claimed that even his critics had given the book a generally positive evaluation. The establishment— Ostrovityanov, director P. A. Khromov and deputy director Maslennikov of the Institute of Economics, Academician I. A. Trakhtenberg, Academician S. G. Strumilin, and others—clearly sided with Varga, and the conservatives made little headway. During most of 1947 the criticism that appeared in journals also seemed mild and cautious, reflecting the absence of official backing for attacks on Varga. For example, even I. A. Gladkov, later one of Varga's most vicious assailants, was rather civil in his article of 15 September 1947 (*Bolshevik,* no. 17, 1947), in which he challenged the book's idea that capitalist states could cure the ills of society by "planning" (for full employment, etc.) and attacked the notion that the state had "acquired decisive significance in the war economy" in the West.[53]

[52] Ostrovityanov did assail Varga for asserting that the struggle between capitalist and socialist systems had halted during the war and also criticized his view of the role of capitalist governments in their economies during the war.

[53] A record of Gladkov's remarkably hard-line speech at a 24 March 1949 meeting of the Institute of Economics (a meeting devoted to fighting cosmopolitans in economics) was later included in the papers that were circulated as an unofficial "Political Diary," and eventually published abroad (*Politicheskiy dnevnik,* vol. 1, pp. 224–35, published by the Alexander Herzen Foundation in Amsterdam in 1972). Viciously attacking Varga and others as cosmopolitans and as pro-Western, Gladkov demanded that they be purged, not debated with.

There is some evidence that suggests that Stalin himself shared some of Varga's views in 1946–47. In a 9 April 1947 interview with Harold Stassen, Stalin expressed optimism about coexistence and economic collaboration between the Soviet Union and the capitalist world. He asked Stassen about the United States government's regulation of the economy and the possibility of preventing an economic crisis. Stalin's comments suggested that he viewed the American economy as strong and unlikely to fall into serious crisis in the near future. Stalin's talk with Stassen was published in *Pravda* on 8 May 1947—the day after the first session of the Varga debate.[54]

Later in 1947, however, Varga's statements on economic planning in the West were seized upon to launch a campaign that clearly enjoyed the political support of the highest circles. The campaign was preceded by a reorganization that stripped Varga of his most influential posts. In October 1947 Varga's institute was merged with the Institute of Economics and both Varga and Khromov, the director of the other institute, lost their jobs, the latter being replaced by Ostrovityanov, deputy academic secretary of the Academy of Sciences' economics and law division.[55] Varga's journal *Mirovoye khozyaystvo i mirovaya politika,* the organ of the Institute of World Economics and World Politics, ceased publication after the December 1947 issue and was replaced by the new journal *Voprosy ekonomiki,* edited by Ostrovityanov.[56]

Voznesenskiy, a Politburo member and deputy premier, clearly bore the primary responsibility for these initial moves against Varga; first because Voznesenskiy was the Politburo's economic theorist

---

[54] The debate on Varga's book was held in three sessions, on 7, 14, and 21 May 1947. Varga may well have had a longtime personal tie with Stalin. According to Abdurakhman Avtorkhanov (*Stalin and the Soviet Communist Party* [New York, 1959], pp. 103–4), Varga had earlier worked in Stalin's private secretariat as head of its foreign section.

[55] The merger of the institutes and appointment of Ostrovityanov were announced in the Academy of Sciences *Vestnik,* no. 12, 1947. V. V. Oreshkin in *Izvestiya akademii nauk SSSR, seriya ekonomicheskaya* ([Economic series of the USSR Academy of Sciences], no. 5, 1975, p. 104) gives 4 October 1947 as the date on which the Presidium of the Academy of Sciences ordered the merger of the institutes.

[56] *Mirovoye khozyaystvo i mirovaya politika* was issued by the new Institute of Economics after the October 1947 merger, but Varga continued as chief editor. The first issue of *Voprosy ekonomiki* was issued in March 1948. Of the seven members of the editorial board of *Mirovoye khozyaystvo i mirovaya politika,* only Varga and Institute of Economics deputy director V. A. Maslennikov were carried over to the editorial board of the new journal. The other five members of the board of *Voprosy ekonomiki* (chief editor Ostrovityanov, I. A. Gladkov, V. P. Dyachenko, G. A. Kozlov, and A. I. Pashkov) were relatively conservative.

and apparently the only Politburo member interested in intervening in this subject, second because he took over Varga's institute in late 1947, and third because he implicitly attacked Varga's views in his own book, published in late 1947. Zhdanov appeared uninvolved,[57] though his son Yuriy, who as head of Agitprop's Science Section was responsible for the field of economic science, presumably sanctioned the attacks on Varga. Reflecting the fact that the attacks on Varga were not considered incompatible with defense of moderates in philosophy and biology, a March 1948 *Vestnik* of the Academy of Sciences (signed to press 20 March) carried an editorial on the three current debates which sided with Kedrov and the moderates in philosophy and biology, but simultaneously joined the attacks on Varga. This editorial presumably reflected Zhdanov's position, resisting the hard-liners in philosophy and biology, but going along with Voznesenskiy's attack on moderates in economics.

The October 1947 reorganization of economic institutes was initiated by Voznesenskiy and its anti-Varga intent became obvious with the December publication of Voznesenskiy's book on the war economy of the Soviet Union, which contradicted Varga's views.[58] According to Voznesenskiy's assistant, Voznesenskiy had asked the CC and Council of Ministers to place the new Institute of Economics under Gosplan's (his) jurisdiction, and once this was done in the fall of 1947, he was able to reorient the institute to serve his needs.[59] In his book, Voznesenskiy, though not attacking Varga by name, dismissed the talk of planning in the West as a "mere wish" and assailed as "sheer nonsense" the views of "some theoreticians who claim to be Marxists" about the "decisive role of the state in the war economy of capitalist countries." He insisted that monopolies, not the state, still controlled policy and that the transition from war to peace would produce crises and mass unemployment in the West. Voznesenskiy's statements were clearly aimed at Varga; in fact, one of the chapters of Varga's book had been printed in *Mirovoye khozyaystvo i mirovaya politika* (no. 1, 1945) under the title, "The

---

[57] Though his mouthpiece *Kultura i zhizn* led the attacks in culture, it carried relatively little in the way of attacks on deviations in economics.

[58] *Voyennaya ekonomika SSSR v period otechestvennoy voyny* [The war economy of the USSR in the period of the fatherland war].

[59] V. V. Kolotov in *Znamya*, June 1974, pp. 133–34. In *Izvestiya akademii nauk SSSR, seriya ekonomicheskaya*, no. 5, 1975, p. 104. Oreshkin writes that after the October 1947 merger of the institutes, "general leadership of scientific work of the Institute of Economics was assigned to Gosplan."

decisive role of the state in the war economy of capitalist countries."
Moreover, during an October 1948 debate at the Institute of Economics, M. L. Bokshitskiy criticized Varga's speech by saying that Varga had not told what conclusions he had drawn from the comments in Voznesenskiy's book about Varga's "non-Marxist concepts."[60]

Immediately upon publication in December 1947, Voznesenskiy's book was advertised by the press as the authoritative word on economics, and Varga came under fire in a series of meetings and press articles, both for his book and for his 1947 statements. Those who had sided with Varga or had written books with similar content were also assailed, and both the old Institute of World Economics and World Politics and the old Institute of Economics were attacked sharply for mistaken liberal trends.

Voznesenskiy's book was lauded in reviews by I. D. Laptev in the 31 December 1947 *Kultura i zhizn* and by Ostrovityanov in the 3 January 1948 *Pravda*. In *Voprosy ekonomiki* (no. 1, March 1948), Ostrovityanov repeatedly cited theoretical propositions laid down in Voznesenskiy's book in his own article on planning, and an unsigned review quoted the book's reassertion that the crisis of capitalism was growing and its rejection of the theses that Western governments played a decisive role in their economies and used elements of planning. Ostrovityanov reviewed the book in *Planovoye khozyaystvo* (no. 1, 1948), and cited it as the example for books on the Soviet economy at a May 1948 meeting of the Institute of Economics.[61]

The first big political attack on Varga appears to have been in the 26 January 1948 *Pravda,* followed on the next day by a meeting at the Institute of Economics to discuss the Varga errors. There followed the attacks on Varga and his colleague I. A. Trakhtenberg in several journal articles and at a 29 March meeting of the Institute of Economics, which was held to discuss a book edited by Trakhtenberg.[62]

Responding to publication of the stenographic record of the May 1947 debate in the November issue of Varga's journal, the conserva-

[60] *Voprosy ekonomiki,* no. 9, 1948, p. 111.
[61] Ibid., no. 4, 1948, pp. 93–94.
[62] *Planovoye khozyaystvo,* no. 1, 1948, signed to press 6 March; *Bolshevik,* no. 5, 1948, dated 15 March; *Voprosy ekonomiki,* no. 1, 1948, signed to press 29 March; and an editorial in the March 1948 Academy of Sciences *Vestnik.*

tive hatchet man I. D. Laptev[63] criticized Varga's "clearly non-Marxist" ideas, especially his claims that capitalist states had acquired a decisive role in their economies (*Pravda*, 26 January 1948). This idea, he said, can lead to the notion that capitalist states can overcome the defects of capitalism and avoid crises by becoming planned societies. At the 27 January 1948 meeting of the Institute of Economics, the institute's new director, Ostrovityanov, attacked the work of the former Institute of Economics and Institute of World Economics and World Politics for "anti-Marxist errors" in treatment of Western economies. Referring to the 26 January *Pravda* attack on Varga's institute and Varga's book, Ostrovityanov criticized Varga for ignoring both the crisis of capitalism and "the struggle between the two systems" and for arguing that the West was moving toward socialism and adopting planning. He noted that Varga had repeated and compounded his reformist views in the May debate and in subsequent articles in the October and November 1947 issues of his journal.

Attacks on Varga and others from his institute soon multiplied. Writing in *Planovoye khozyaystvo* (no. 1, signed to press 6 March 1948), I. N. Dvorkin assailed the idea of "planned" capitalism and criticized Varga and his colleagues, giving special attention to a 1947 book edited by Trakhtenberg, *Voyennoye khozyaystvo kapitalisticheskikh stran i perekhod k mirnoy ekonomike* (The war economy of capitalist countries and the transition to a peacetime economy). At a 29–30 March 1948 meeting of the Institute of Economics, deputy director G. A. Kozlov attacked the Trakhtenberg book for having "the imprint of bourgeois ideology" (*Voprosy ekonomiki*, no. 2, 1948, p. 107), and similar criticisms were made by Ostrovityanov and others. In *Bolshevik*, no. 5, L. M. Gatovskiy also attacked the Trakhtenberg book and criticized the Institute of World Economics and World Politics for having issued recently "a number of non-Marxist" books about capitalist economies. In the same issue, I. I.

[63] Laptev, a Lysenko ally, had assailed Lysenko foes A. R. Zhebrak and N. P. Dubinin in a vicious 2 September 1947 *Pravda* article and was appointed to VAS-KhN IL in July 1948 as part of the Lysenkoist influx. Later, after the late 1952 renewal of the attack on Voznesenskiy, Laptev had to confess having erred in his earlier praise of Voznesenskiy's "anti-Marxist" book (*Pravda*, 16 January 1953). He eventually became director of the Institute of Economics, but was ousted in 1958 for siding with Khrushchev's foes and opposing Khrushchev's agricultural reforms. In early 1962 he was expelled from VASKhN IL after bitter attacks on him by Khrushchev at the March 1962 CC plenum.

Kuzminov, the deputy chief editor of *Bolshevik,* cited Voznesenskiy's refutation of the idea of "some of our 'theoreticians' about the decisive role of the bourgeois state in the economy" and assailed Varga for asserting the existence of "planned bases" in the economies of capitalist countries. He also criticized Trakhtenberg's speech at the May 1947 debate for urging a favorable attitude toward American reforms in planning. A 19 April Academy of Sciences meeting heard Ostrovityanov assail Varga and his institute for a "harmful—reformist—orientation."[64] Ostrovityanov declared that Varga had assumed that the struggle between socialism and capitalism had stopped during World War II and had argued that capitalist states were adopting elements of socialism and becoming more like the Soviet Union. At an October 1948 meeting of the Institute of Economics, others attacked the view that the struggle between the two systems had abated during the war, and L. Ya. Eventov admitted that such views were widespread in Varga's old institute.[65]

In the face of these attacks, Varga began a retreat, but only slowly and grudgingly. At the Institute of Economics' October 1948 meeting, he conceded that at least some of his views were mistaken or had been misunderstood, and his colleagues V. A. Maslennikov, L. Ya. Eventov, and M. L. Bokshitskiy admitted their own errors in arguing similar views.[66] However, Varga still defended many of his controversial views; in fact, as Ostrovityanov complained, he "persisted in denying his most gross principled mistakes." Ostrovityanov warned him that "you should know from the history of our party what sad consequences stubborn insistence in one's mistakes leads to."[67]

Varga's next move came only months later, when, in a letter published in the 15 March 1949 *Pravda,* he sought to prove that he had no pro-Western sympathies but was patriotic and politically orthodox. He assailed the Western press for portraying him as "a 'person with a Western orientation,' a 'defender' of the Marshall Plan, who denies the possibility of a crisis of overproduction in the U.S." He declared that he had been the first scholar in the Soviet Union publicly to attack the Marshall Plan (in a 27 August 1947 lecture) and argued not only that he had not denied the possibility of

[64] *Vestnik* of the Academy of Sciences, June 1948, p. 73.
[65] *Voprosy ekonomiki,* no. 9, 1948, p. 76.
[66] Ibid., nos. 8 and 9, 1948.
[67] Ibid., no. 9, 1948, p. 96.

an economic crisis in the United States, but that he even had predicted the onset of such a crisis by 1948—a prediction that he claimed had turned out to be correct. In effect, Varga's letter was a defense of his views, for it simply denounced the West for slandering him and made no mention of error in any of his positions. A few days later, however, Varga was finally forced to recant. At a "late March" meeting of the Institute of Economics, reported in the March 1949 *Voprosy ekonomiki* (actually signed to press only on 16 April), Varga, Trakhtenberg, Eventov, Bokshitskiy, and other former workers at Varga's old institute were accused of "errors of a cosmopolitan character" and "bowing to things foreign" (*inostranshchina*), and Varga and Trakhtenberg reportedly admitted committing errors. This same issue of *Voprosy ekonomiki* carried a ten-page article by Varga admitting that his book—as well as other books produced by his institute ("for which I bear responsibility")— had been correctly criticized for bearing a "reformist orientation." Moreover, he declared that his book contained "errors of a cosmopolitan orientation," since it "whitewashed capitalism." He also admitted that he had erred in not immediately acknowledging the correctness of the criticisms, but, he wrote, "better late than never."

By the time Varga was finally forced to renounce his views, however, the political situation had changed: his chief assailant Voznesenskiy was dismissed in early March, and the pressure to follow up on moves against Varga apparently slackened. Economists were still divided on Varga, but though the extremists pressed for harsher action, the establishment settled for his recantation. This division was evident in the March debate at the Institute of Economics. *Voprosy ekonomiki* reported that after Varga and others had admitted errors, some speakers pointed to the "insufficiency and partial character of these admissions and demanded that they criticize their errors in the press and produce works permeated with the spirit of Soviet patriotism." According to a later *samizdat* account, I. A. Gladkov, one of the extremists, assailed Varga and all his colleagues as "rootless cosmopolitans" and demanded that they be purged from the leadership of both the institute and *Voprosy ekonomiki*.[68] In contrast, Ostrovityanov's closing speech, as reported in *Voprosy ekonomiki*, sounded conciliatory. The institute's

---

[68] See the account of Gladkov's speech at the 24 March 1949 meeting of the Institute of Economics reported in *Politicheskiy dnevnik*, vol. 1, pp. 224–35.

director declared that only an "insignificant group" of economists was actually "infected" with cosmopolitanism and that "one should distinguish between rootless cosmopolitans and people who have made certain mistakes in their work." Clearly, Ostrovityanov did not consider Varga a hopeless "rootless cosmopolitan," but rather someone who had erred, could correct his errors and continue his work. Apparently as a result, Varga remained on the editorial board of *Voprosy ekonomiki* through Stalin's death and continued to write articles. It is another sign of the rapid cooling-off of the campaign that no further accounts of the March meeting and its attacks on Varga and the cosmopolitans ever appeared in *Voprosy ekonomiki*, despite the March journal's promise to print a fuller account in "its next issues."

Hence, Varga, despite the long campaign against him, got off with a public acknowledgment of errors—in contrast to his assailant Voznesenskiy, who was arrested and executed. Varga survived and even participated in the late 1952–early 1953 debates on Voznesenskiy's errors, after Voznesenskiy was condemned publicly for holding mistaken views.[69]

Thus, during 1947 the atmosphere in these fields changed noticeably from toleration of a range of views to persecution by extremists of anyone who contradicted the new orthodoxies. This trend clearly paralleled the development of the cold war and establishment of the "two camps" thesis. Although Zhdanov himself had the role of enunciator of this new hard line both in philosophy and in foreign affairs (in his June and September 1947 speeches), he did not gain politically as he had in the 1946 crackdown in literature and culture. Whether or not he initiated or even agreed with this shift during 1947, he lost ground politically as the moderates, who were more or less associated with him, were attacked as an unpatriotic and politically unreliable element in a world divided into two hostile camps. At the same time, the position of dogmatists, who were associated with Zhdanov's foes, became stronger. All of these developments prepared the ground for the final intrigues which would bring about Zhdanov's downfall.

[69] See the 7–10 January 1953 discussion on Stalin's economic theses and Voznesenskiy's errors held in the economics and law division of the Academy of Sciences and the Institute of Economics (*Voprosy ekonomiki*, no. 4, 1953).

CHAPTER 3

Zhdanov's Fall — 1948

The immediate cause for Zhdanov's fall appears to have been his stand with the moderates in philosophy and science, especially against Lysenko, combined with the break with Yugoslavia. Zhdanov's fall in mid-1948 coincided exactly with the unfavorable outcome of the Lysenko debate and the eruption of the conflict with Tito. Zhdanov's decline became noticeable early in 1948; he disappeared from the scene in July, and died of heart failure at the end of August.

The triumph of dogmatism and chauvinism in science and philosophy, as well as in foreign policy, brought Zhdanov's rivals Malenkov, Beriya, and Suslov to the fore. Whereas Lysenko, Yudin, Mitin, and Suslov clearly held ultraconservative viewpoints, it is not clear whether Malenkov or Beriya shared these ideas or simply used the resurgence of dogmatism to increase their power and destroy Zhdanov's faction. Nor is it clear exactly what role Beriya and Malenkov played in the 1947–48 debate in science and philosophy or even in the dispute with Tito — although Beriya later was accused of provoking the split with Yugoslavia.

In any case, Malenkov, Beriya, and Suslov immediately began to root out Zhdanov's influence. Malenkov, once more in control of the CC apparat, reversed Zhdanov's 1946 reorganization of the cadre organs; Suslov, in charge of ideology, reorganized and purged Zhdanov's Agitprop; and Beriya began fabricating the Leningrad Case, which in 1949 brought the second round of Zhdanovite ousters.

Zhdanov and Lysenko

Zhdanov's fall was connected with Stalin's support for Lysenko's drive to take over Soviet biology. All available evidence suggests that Zhdanov was, at the least, cool toward Lysenko; more directly, his son Yuriy publicly acknowledged having attacked Lysenko and having defended the geneticists. The date of Lysenko's triumph also corresponded more closely than any other event with Zhdanov's final fall.

Although most outside observers have overlooked the evidence for and consequences of Zhdanov's anti-Lysenko position, Soviet biologist Zhores Medvedev directly contends that Zhdanov resisted Lysenko's rise and that this was virtually the major cause of Zhdanov's fall. According to Medvedev, Lysenko sent a long memo to Zhdanov in mid-1947, asking the latter to support his viewpoint. This appeal initiated a high level study of the situation in biology.[1] However, instead of backing Lysenko, Zhdanov, according to Medvedev, attacked Lysenko at Orgburo meetings in the spring of 1948 and proposed removing him as president of VASKhNIL.[2] One careful student of Soviet science, Loren Graham, in examining issues of philosophy of science, also concluded that Zhdanov had opposed both Lysenko personally and the party's intervention in science on his behalf. While noting Zhdanov's hard line in culture, Graham declares that "until the month of Zhdanov's death, however, natural scientists escaped the rule by decree that obtained in other cultural fields."[3] He reports that he was unable to find one mention of Lysenko in any of Zhdanov's speeches and that Lysenko did not even try to portray Zhdanov as a defender of Lysenkoism.[4]

Although Zhdanov himself did not publicly go on record against Lysenko, his son Yuriy did. The twenty-eight-year-old Yuriy, a chemist by training, became head of Agitprop's Science Section sometime in 1947, although he was first identified as head of the section only in the 22 April 1948 *Pravda*.[5] In this post, he apparently

---

[1] Medvedev, *The Rise and Fall of T. D. Lysenko*, pp. 112–16.

[2] *Grani*, no. 71, May 1969, p. 98. Medvedev also claims that the Leningrad party organization had backed Lysenko's foe N. I. Vavilov in 1940 when Lysenko, as VASKhNIL president, was attempting to interfere in Vavilov's Leningrad Plant Growing Institute (Medvedev, *The Rise and Fall of T. D. Lysenko*, pp. 149–50).

[3] Loren Graham, *Science and Philosophy in the Soviet Union* (New York, 1972), p. 19.

[4] Ibid., p. 444.

[5] The *Bolshaya sovetskaya entsiklopediya* yearbook for 1971 lists him as leaving

supervised the work of *Voprosy filosofii* and obviously must have sanctioned the unorthodox articles published by Kedrov, as well as the resistance to Lysenko by Soviet biologists. Furthermore, he made a veiled attack on Lysenko in a *Komsomolskaya pravda* article (25 May 1948), in which he criticized those scientists who used their positions and authority to block development of other schools of thought and tried thereby to create a "personal monopoly" in their particular field. His most fateful act, however, was to attack Lysenko directly in the April 1948 speech "Controversial questions of present-day Darwinism," delivered at a seminar of local party lecturers. Although the *Kultura i zhizn* account of the seminar (11 April 1948) did not report his comments, Yuriy admitted in a 10 July letter of recantation that he had sharply attacked Lysenko.[6] Mitin, as deputy chairman of the All-Union Society for Dissemination of Political and Scientific Knowledge, had helped organize the seminar and was present. According to Medvedev, Mitin told Lysenko about Yuriy's attack, and Lysenko and Mitin then went to Stalin and complained of intrigues by Zhdanov and his son. Medvedev claims that Stalin, who had already cooled toward Zhdanov because of the latter's growing popularity and power and because of anti-Zhdanov intrigues by Malenkov and Beriya, seized on this information to denounce Zhdanov sharply and to force his son to make a public confession of error. In a letter to Stalin dated 10 July (but only published in the 7 August *Pravda* — during the notorious August 1948 session of VASKhNIL), Yuriy Zhdanov wrote that since recently receiving his new post in the CC's Science Section, he had listened to Lysenko's scientific foes and had taken their recommendations "under my protection." He wrote that "speaking at a seminar of lecturers on controversial questions of contemporary Darwinism," he had erred, that "I underestimated my responsibility and had not imagined that my speech would be regarded as the official point of view of the CC." "My sharp and public criticism of Academician Lysenko was a mistake," he concluded. Stalin's daughter Svetlana Alliluyeva, who was Yuriy's wife from 1949 until 1953, remarked that in 1948 he had come out against Lysenko but that Stalin had instantly defended the scientist. She reports that Yuriy

---

Moscow State University and going into unspecified party work in 1947. Medvedev declares that he took over the Science Section immediately upon leaving the university (Medvedev in *Grani*, no. 71, May 1969, p. 101).

[6] The letter was published in the 7 August 1948 *Pravda*.

then said, "now genetics is finished," and that he had to write his letter of repentance and refute the chromosome theory.[7]

After his son's letter, the elder Zhdanov went on leave.[8] Stalin secretly endorsed Lysenko's stand in biology and, bypassing elections, on 15 July simply appointed thirty-five new members of VASKhNIL, including many Lysenkoists.[9] At an August 1948 session of VASKhNIL, Lysenko crushed his foes by luring them into public attacks on his positions and then later in the session revealing that Stalin himself had already endorsed these positions.[10] The elder Zhdanov died shortly afterward—on 31 August 1948—and Lysenko quickly became more dominant in science than ever before, either purging his foes or forcing them to accept his views.

Stalin's ire was directed at Zhdanov senior, despite the fact that it was Yuriy Zhdanov who had publicly attacked Lysenko. Thus, Yuriy was still granted the degree of candidate of philosophy on 29 June, after his father had apparently already gone into seclusion.[11] Moreover, despite Yuriy's public humiliation in July-August 1948, his articles continued to appear in the press,[12] and, though virtually all Zhdanov men in Agitprop and the press were ousted in the year after Zhdanov's death, Yuriy retained his post as head of the CC Science Section.[13]

One reason for Yuriy's retention may have been his marriage to Stalin's daughter Svetlana. Khrushchev recalled that sometime after the war Stalin had forced his daughter to divorce her husband, who was Jewish, and marry Yuriy Zhdanov.[14] Svetlana, in her book *Twenty Letters to a Friend*, denies that Stalin forced her to marry

---

[7] Alliluyeva, *Only One Year,* p. 380. Yuriy attacked Kedrov and lauded Lysenko in a November 1951 *Bolshevik* (no. 21) article.

[8] Medvedev in *Grani,* no. 71, May 1969, pp. 98–102.

[9] Medvedev, *The Rise and Fall of T. D. Lysenko,* pp. 115–16. The 15 July decree of the Council of Ministers which announced the appointments was published in the 28 July 1948 *Pravda,* on the eve of the August session of VASKhNIL.

[10] Later, in an 8 March 1953 *Pravda* article, Lysenko stated that Stalin personally had edited his August 1948 VASKhNIL speech.

[11] On 1 July 1948, *Pravda* announced that on 29 June the Institute of Philosophy's scholarly council had unanimously accepted Yuriy's dissertation and awarded him the degree of candidate of philosophy.

[12] In the 21 September 1949 and 21 December 1950 *Kultura i zhizn,* and February 1951 (no. 3) and November 1951 (no. 21) issues of *Bolshevik.*

[13] Yuriy was still identified as head of the section in the 16 October 1951 *Sovetskaya Belorussiya.* After the 1950 Agitprop reorganization, it had been raised to the status of an independent section, rather than just a section within Agitprop.

[14] *Khrushchev Remembers,* pp. 292–93.

Yuriy but says that he had hoped she would. She married him in the spring of 1949, but divorced him shortly before Stalin's death, thereby, she said, displeasing her father.[15] Despite his moves against Yuriy's father, then, Stalin appears to have held nothing against Yuriy Zhdanov.

The Break with Yugoslavia

One of the key factors in Zhdanov's fall must have been the break with Tito, since Zhdanov played the leading role in the Cominform and relations with Yugoslavia and since the final break coincided with Zhdanov's loss of influence. In fact, the 28 June 1948 Cominform meeting that formally excommunicated Tito also happened to be Zhdanov's last public appearance. Moreover, the other key actors in the Cominform-Tito drama were Zhdanov's foes on the domestic scene—Malenkov, Suslov, and Yudin. In addition, Khrushchev later accused Beriya and MGB chief V. S. Abakumov of having played a "provocative role" in the split with Tito[16]—although no evidence was provided to support this accusation. Whether Zhdanov had friendlier relations with Yugoslavia than his rivals is not clear, although his Leningrad protégés were later accused of cooperating perhaps too enthusiastically with Milovan Djilas on his January 1948 visit to Leningrad. In any case, Zhdanov was intimately involved in the Cominform intrigues that ended with Tito's successful defiance of Stalin.

The Cominform was created in September 1947 with Zhdanov and Malenkov present to represent the USSR. Zhdanov gave a major speech at the opening meeting and clearly played the key role. For example, according to Tito's spokesman Vladimir Dedijer, Zhdanov criticized the French and Italian communist parties and instructed the Yugoslav delegates to attack them as well.[17] Malenkov also gave a speech, but it focused mainly on Soviet domestic affairs and was published only months later—in the 9 December 1947 *Pravda*.

By early 1948 Stalin clearly was becoming increasingly suspicious of Tito and his independent attitudes, and Soviet leaders deliberately began to provoke a split. There were signs that relations between Yugoslavia and the USSR were already seriously strained by late

---

[15] *Twenty Letters to a Friend* (New York, 1967), pp. 192, 197, 210–11.

[16] In his speech upon arriving in Yugoslavia on 26 May 1955 (*Pravda*, 27 May 1955).

[17] Vladimir Dedijer, *Tito Speaks* (London, 1953), p. 304.

January 1948. On 28 January 1948, *Pravda* attacked the idea of a Balkan Federation, or customs union, which had been mentioned by Bulgarian leader Georgi Dimitrov. Meanwhile, a delegation of Yugoslavs under Djilas was in the midst of difficult negotiations with Stalin and at the final session on 10 February 1948 Stalin sharply argued with the Yugoslav and Bulgarian delegates, notably over alliances between Balkan states.[18] Dedijer terms this session "the beginning of Stalin's open pressure on Yugoslavia" and notes that Stalin ordered Tito's portraits taken down in Rumania right after 10 February.[19] During February Soviet trade officials, then engaged in negotiating a new trade treaty with Yugoslavia, told the Yugoslav negotiators in effect that there would be no new trade treaty.[20] By 1 March Tito was telling a Yugoslav CC plenum that relations with the Soviet Union were at an impasse,[21] and on 18 and 19 March Soviet officials informed the Yugoslavs that all Soviet military and civilian advisers were being withdrawn from Yugoslavia because they were "surrounded by hostility." Tito responded with a 20 March letter to Molotov protesting the withdrawal and the charge of Yugoslav hostility.[22] On 27 March, a Soviet letter to Tito accused the Yugoslav police of following Soviet representatives in Yugoslavia, specifically mentioning Soviet Cominform representative P. F. Yudin. The Yugoslavs denied the Soviet charges, including that of spying on Soviet representatives, in their 13 April reply. On 16 April Tito received a letter signed by Zhdanov pressing the split,[23] and in a 4 May letter Soviet leaders declared that they chose to believe their own representatives in Yugoslavia, "among them comrade Yudin," rather than Tito or Kardelj. In their 17 May reply, the Yugoslavs declared that the 4 May Soviet letter, with its uncompromising rejection of Yugoslavia's explanations, made it clear that the USSR wanted no reconciliation, and the Yugoslavs then declared that they would not attend the June Cominform meeting, which was sched-

---

[18] See Dedijer's description, ibid., pp. 325–32, and Milovan Djilas's description in his *Conversations with Stalin* (New York, 1962), pp. 133–77. Molotov, Zhdanov, Malenkov, and Suslov also participated.

[19] Dedijer, *Tito Speaks,* p. 334.

[20] Ibid., p. 335.

[21] Ibid., p. 335.

[22] The 20 March letter, as well as subsequent 27 March, 13 April, 4 May, and 17 May letters between the Soviets and the Yugoslavs, is published in *The Soviet-Yugoslav Dispute* (London: Royal Institute of International Affairs, 1948).

[23] Dedijer, *Tito Speaks,* p. 358.

uled to discuss the Soviet-Yugoslav dispute. On 19 May, a letter signed by Suslov urged the Yugoslavs to attend the June meeting, but Tito refused.[24] The Cominform went ahead with its meeting on 28 June without Yugoslav participation, adopting a resolution that declared the Yugoslavs outside the Cominform and then making public the correspondence between the two parties—an act that made the break public and final. At that 28 June meeting, the USSR was represented by Suslov, in addition to Zhdanov and Malenkov, a fact that suggests Zhdanov's diminished influence. In the all-out campaign that later developed against Yugoslavia, Suslov and Yudin played the leading roles, representing the USSR at the November 1949 Cominform meeting, for example.

Although there were naturally much more basic reasons for the split, one of the most immediate causes—or pretexts—was the series of reports from Soviet representatives in Yugoslavia about harassment on the part of the Yugoslavs and about hostility toward the USSR—reports denied by the Yugoslavs. It remains unclear whether Stalin encouraged Soviet representatives to send such anti-Yugoslav reports and complaints because he was already bent on attacking Tito and needed a pretext, or whether the Soviet representatives initiated these complaints on their own, feeding Stalin's suspicions and helping turn him against Tito. What is clear is that the main source of these complaints against the Yugoslavs was the Soviet representative in the Cominform, P. F. Yudin. Yudin had come to Belgrade as the Soviet representative and the chief editor of the Cominform's organ when the Cominform was headquartered in the Yugoslav capital late in 1947. According to Dedijer, Yudin completely controlled the paper and his high-handed activities immediately began causing friction with the Yugoslavs. In addition, according to Dedijer, he "tried hard to poison relations between Yugoslavia and her neighbors."[25] Furthermore, he was soon complaining of being spied on by Yugoslav police, for such charges appear in the 27 March 1948 Soviet letter to the Yugoslavs. Yudin was singled out in the various Soviet letters as the main victim of Yugoslav harassment, a fact that suggests his key role in aggravating Soviet-Yugoslav relations. In his recollections Khrushchev treats Yudin as a primary cause of the quarrel with Tito, and even of the

---

[24] Ibid., p. 365.
[25] Ibid., pp. 306–8. Dedijer also describes Yudin's control of the paper in his book, *The Battle Stalin Lost* (New York, 1971), pp. 119–20.

later quarrel with Mao.[26] Khrushchev relates how Yudin informed Stalin that at some party meeting the Yugoslavs made "all sorts of sarcastic, disrespectful, and even insulting remarks about the Soviet Union and particularly about our military and technical advisers." Stalin ordered that copies of Yudin's report be sent to all Politburo members, and shortly thereafter, the USSR ordered all its advisers home, thereby bringing the situation to a head.[27] Khrushchev notes that later, as ambassador to China, Yudin reported anti-Soviet statements by Mao and clashed with Mao over philosophy.[28]

Yudin also played a direct role in the 1948 exchanges between Moscow and Tito. As Tito later told Dedijer, Yudin had visited the Yugoslav leader a week before the first Soviet letter to sound out Tito's reactions. Later, Yudin personally presented Tito with Hungarian leader Rakosi's 15 April letter attacking the Yugoslavs. Yudin also played a role in trying to lure Tito to the June Cominform meeting.[29]

Zhdanov could hardly have been enthusiastic about the choice of Yudin to play a key role in the Cominform. Zhdanov had apparently been instrumental in stripping Yudin of most of his leading posts from 1944 to 1946, and Yudin's hostility to Zhdanov was evident during the June 1947 philosophy debate when he bitterly assailed Zhdanov's protégés. Yudin presumably owed his selection to Stalin's favor and must have looked to Malenkov and Suslov for patronage. Even though Yudin's negative reports on the Yugoslavs may have stemmed from a real mistrust of Yugoslav independence, they probably were also aimed at undermining his personal enemy Zhdanov. In contributing to Stalin's hostility toward Tito, Yudin probably was deepening the Soviet leader's distrust of Zhdanov as well.

Zhdanov's Final Demise

Public signs point to July as the month of Zhdanov's final fall from favor. His last public appearance was at the 28 June Cominform meeting that excommunicated Yugoslavia, apparently a drastic setback for Zhdanov, who had been in charge of Cominform affairs.

[26] *Khrushchev Remembers*, p. 464.
[27] Ibid., p. 376.
[28] Ibid., pp. 464–65.
[29] These three episodes are described in Dedijer's *The Battle Stalin Lost*, pp. 108, 122, 125.

On 10 July, his son wrote his letter recanting his opposition to Lysenko, clearly a humiliating defeat for Zhdanov. Zhdanov's defeat was confirmed when *Pravda* on 21 July 1948 published a 20 July CC message to the Japanese Communist party signed by Malenkov as CC secretary. Since Malenkov's fall in late 1946, Zhdanov had had the privilege of signing messages or decrees as Stalin's top deputy (for example, a 14 December 1947 decree). Malenkov's signature on the 20 July message revealed not only that he had been restored to his key post of CC secretary, but that he had replaced Zhdanov as Stalin's top deputy in the party.

Zhdanov had slipped in status prior to July 1948, but still ranked as Stalin's top deputy in the party and still played a crucial, if somewhat diminishing, role. The first sign of slippage was his loss of control over Agitprop and the press in late 1947, when Suslov became Agitprop chief and CC secretary. The presence of Zhdanov's foes at important functions in Zhdanov's spheres of foreign affairs and ideology, even though Zhdanov still played the main role, was another indication of his declining power. Malenkov, as well as Zhdanov, gave a speech at the September 1947 meeting that established the Cominform, and in mid-January 1948 Suslov attended the CC conference on music which laid down the law to composers. Although Zhdanov delivered the main address at the music conference and apparently took the lead in drafting the 10 February 1948 CC decree on Muradeli's opera,[30] his music speech—in contrast to his earlier speeches on culture—was not published in *Pravda*, another possible sign of diminishing influence.

The role of Suslov, the new CC secretary, soon expanded. He delivered the annual Lenin anniversary speech on 21 January 1948, and by March he was running ideological functions. Zhdanov's public appearances became rare. This decline was probably also connected with the rapid deterioration of Zhdanov's health. According to Khrushchev, Zhdanov had long been in bad health and, particularly in the last months before his death, he could not control

---

[30] *Kultura i zhizn*, 11 February and 21 March 1948. Alexander Werth (*Russia: The Post-War Years*, pp. 350–73) gives a detailed account of the January meeting on music and based on the published verbatim report declares that Zhdanov dominated the meeting.

his drinking.[31] Stalin, who usually encouraged others to drink heavily, began to shout frequently at Zhdanov to stop drinking, according to Khrushchev. During his January-February 1948 visit, Djilas noted that Zhdanov was the only Politburo member who drank orangeade instead of vodka, and attributed this to Zhdanov's heart trouble.[32]

Still, Zhdanov continued to rank third after Stalin and Molotov, whereas Malenkov usually ranked far down the line. In the 22 December 1947 *Pravda* editorial, Zhdanov was third in the list of candidates for election to the local soviets. He retained that position in lineups at the 21 January 1948 anniversary of Lenin's death (*Pravda*, 22 January), in the list of delegates elected to the Moscow soviet in the 25 January *Pravda*, in the lists of Stalin's dinners given in the 27 January, 6 February, 21 February, 20 March, and 8 April 1948 *Pravdas*, and in lineups at the USSR Supreme Soviet session (1 February *Pravda*) and at Armed Forces Day (24 February *Pravda*). Malenkov ranked seventh or eighth in late 1947, but rose slightly in early 1948—possibly another early sign of Zhdanov's slippage.[33] On May Day 1948, the date of the last protocol lineup in which Zhdanov appeared, Zhdanov was listed in his customary third position, with Malenkov sixth. In the photo of the tribune itself, however, Zhdanov was standing eighth and Malenkov fifth.[34] The next public lineup of the Politburo was on Air Day, 25 July, but Zhdanov was apparently then too sick to attend. On that occasion, Malenkov stood fourth after Stalin, Beriya, and Voroshilov (Molotov being absent). Zhdanov died on 31 August.

---

[31] *Khrushchev Remembers*, p. 284.

[32] Djilas, *Conversations*, p. 155.

[33] Malenkov was seventh on 7 and 8 November 1947 (he would have been eighth but Zhdanov was absent), eighth in a 16 November 1947 *Pravda* listing, and seventh in the 22 December *Pravda* editorial list. He was fifth at the 21 January 1948 Lenin anniversary, seventh in the 25 January 1948 *Pravda* list of Moscow soviet deputies, fifth in the 27 January *Pravda*, eighth at the Supreme Soviet session reported in the 1 February *Pravda*, sixth at Stalin's dinners reported in the 6 and 21 February *Pravda*s, seventh on Armed Forces Day as cited in the 24 February *Pravda*, fifth in the 20 March *Pravda*, and sixth in the 8 April *Pravda*.

[34] The order of listing was: Stalin, Molotov, Zhdanov, Beriya, Voroshilov, Malenkov, Mikoyan, Kaganovich, Voznesenskiy, Shvernik, Kosygin, Kuznetsov, Suslov, Popov, with Bulganin speaking. The standing in the photo was Voroshilov on Stalin's right and Bulganin on his left, followed by Molotov, Malenkov, Beriya, Kaganovich, Zhdanov, Mikoyan, Shvernik, Voznesenskiy, Suslov, Kosygin, Kuznetsov, Shkiryatov. The listing would appear the more authoritative of the two.

Rise of Beriya, Malenkov, and Suslov

With Zhdanov's death, Beriya and Malenkov became Stalin's most powerful deputies, while new CC Secretary Suslov took over direct responsibility for Zhdanov's functions in ideology and foreign affairs. Beriya strengthened his personal control over the police and began fabricating criminal cases against Zhdanov's vulnerable allies and protégés. Malenkov took over and reorganized the CC cadres apparat and brought reliable protégés into the leadership. Suslov reorganized and purged Zhdanov's erstwhile ideological apparat.

Zhdanov had ranked third after Stalin and Molotov, and this amounted to the position of heir apparent, for although Molotov formally ranked second, he did not have commensurate power. Moreover, after early 1949 Molotov appears to have lost almost all real power, even though he retained his formal position until Stalin's death. On 4 March 1949 Molotov gave up his post as foreign minister, remaining deputy premier and Politburo member. At about the same time, according to Khrushchev and Alliluyeva, Molotov's Jewish wife Polina Zhemchuzhina was arrested as a Zionist and exiled to Kazakhstan. Alliluyeva states that after the 1949 arrest of his wife, Molotov was excluded from power and from Stalin's inner circle and in fact was not even called in when Stalin fell ill in March 1953.[35] Khrushchev describes Stalin's arrest of Molotov's wife as part of Stalin's purge of Jewish leaders in 1948–49 and claims that this act helped to ruin relations between Stalin and Molotov.[36] Khrushchev states that Molotov refused to vote for expelling his wife from the CC at a subsequent CC plenum, infuriating Stalin, who then began abusing Molotov. The plenum in question is presumably the one held after the 1952 party congress, for there were no other plenums between 1947 and 1952. At this plenum, according to Khrushchev, Stalin accused Molotov of being an agent of the United States and, although he included Molotov in the new twenty-five-man Presidium elected at the plenum, he excluded him from the inner Politburo—the nine-man bureau of the Presidium.[37] Khrushchev

---

[35] *Twenty Letters to a Friend*, pp. 109, 196, 208.

[36] *Khrushchev Remembers*, pp. 260–61. Khrushchev relates how, after the war, the Jewish Anti-Fascist Committee of the Sovinformburo asked Stalin to turn the Crimea into a Jewish Autonomous Republic and how Stalin, enfuriated by this, later had the committee's leaders, S. A. Lozovskiy and S. M. Mikhoels, killed. Molotov's wife was somehow involved and was later arrested as well.

[37] Ibid., pp. 281, 308–09.

declared that Stalin thereafter excluded Molotov from meetings and informal gatherings and clearly planned to do away with him.[38]

After Zhdanov's death, the key number three position appeared to alternate between Beriya and Malenkov during 1948 and 1949. At the 7 and 8 November 1948 anniversary ceremonies, with Beriya absent, Malenkov was listed third, but on most occasions in late 1948 and early 1949 Beriya was third. At a 15 December 1948 dinner, Beriya was listed third and Malenkov sixth. At Moscow and Leningrad party conferences reported in *Pravda* on 18, 23, and 24 December 1948, at the Azerbaydzhani, Georgian, Latvian, Kirgiz, and Lithuanian congresses in January and February 1949 (as reported in local papers) and the Leningrad Komsomol conference (*Leningradskaya pravda*, 1 February 1949), Beriya was listed third and Malenkov fourth. However, at the 22 January 1949 Lenin anniversary ceremony, the February 1949 Belorussian congress (*Sovetskaya Belorussiya*, 16 February 1949), the 10 March Supreme Soviet session, a 1 April 1949 dinner, and the 1 May 1949 parade, Malenkov was listed third and Beriya fourth.

The power of the first of Zhdanov's successful rivals, Beriya, rested on the police apparatus: the Ministry of State Security (MGB) and the Ministry of Internal Affairs (MVD). In his postwar intrigues, Beriya relied mainly on the former, especially on its minister, V. S. Abakumov.[39] Although Beriya had personally headed the MVD[40] during the war, the postwar leaders of the MVD were not so close to Beriya or so involved in his crimes. Many of the 1946 MVD leaders survived Beriya and even rose after his 1953 execution,[41] but MGB

---

[38] Ibid., pp. 308–10.

[39] In his recollections, Khrushchev declared Abakumov Beriya's man, loyal to Beriya even ahead of Stalin (ibid., pp. 104, 256, 312).

[40] Called the People's Commissariat of Internal Affairs (NKVD) until early 1946.

[41] S. N. Kruglov, minister from January 1946 to March 1953, succeeded Beriya as minister in July 1953 (when Beriya was arrested) and continued in this post for over two years. Deputy minister I. A. Serov became chairman of the newly established KGB in 1954; deputy minister A. P. Zavenyagin became deputy premier and minister of medium machine building in 1954; and deputy minister V. S. Ryasnoy was still chief of the Moscow *oblast* MVD in 1955 (so identified in the 1 February 1955 *Moskovskaya pravda*). According to émigré Yu. Krotkov, who had close ties with the family of Beriya's deputy V. G. Dekanozov, it was Serov who signed the orders for the arrest of Beriya's closest assistants upon Beriya's arrest in July 1953 (see Krotkov's "Konets Marshala Beriya" [The demise of Marshal Beriya] in the December 1978 *Novyy zhurnal*, pp. 224, 230–31). Khrushchev notes that he trusted Serov, who had been police chief in the Ukraine with him, and that when the Politburo arrested Beriya in mid-1953, he had proposed entrusting Serov rather than any other police official with holding Beriya (*Khrushchev Remembers*, pp. 115, 338).

leaders were not so fortunate. Both V. N. Merkulov, minister until late 1946, and his successor, Abakumov, were later executed for participating in Beriya's plots.[42] That the MGB was Beriya's political instrument was recognized when Stalin moved against Beriya in late 1951. At that time, Stalin had Abakumov arrested, appointed CC official S. D. Ignatyev minister and Odessa First Secretary A. A. Yepishev deputy minister, and had Ignatyev initiate the Doctors' Plot.[43]

Beriya clearly controlled the MGB through Merkulov in 1946, but according to Khrushchev's secret speech, Stalin placed Zhdanov's protégé, CC Secretary Kuznetsov, in charge of state security sometime after Kuznetsov joined the Secretariat in March 1946.[44] There seems no reason to doubt Khrushchev's assertion, but neither is there any public evidence to back it up. Although there were numerous shifts in MGB and MVD leadership in 1946 and 1947, there appears to be no evidence that persons hostile to Beriya were placed in the leadership of these ministries or that Beriya subsequently ousted any of them in 1948 or 1949.[45] However, Beriya's intrigues against Kuznetsov do appear to lend some substance to Khrushchev's statement. As soon as Kuznetsov's patron Zhdanov was dead, Beriya and Abakumov began fabricating the Leningrad Case—aimed precisely against Kuznetsov and his colleagues. In his secret speech Khrushchev claimed that Kuznetsov's rise had alarmed Beriya and that Beriya "had 'suggested' to Stalin" the fabrication of a case against Kuznetsov. Abakumov carried out the fabrication, eventually leading to the execution of Kuznetsov and others.[46]

Beyond the Leningrad Case, however, evidence of Beriya's activities during 1946–49 is scarce. Although press coverage of activity in the party apparat and ideological establishment was adequate to

---

[42] Merkulov was shot with Beriya in December 1953, according to an official announcement in the 24 December 1953 *Pravda*, whereas Abakumov was shot after being convicted (at a December 1954 trial) of fabricating the Leningrad Case (*Pravda*, 24 December 1954).

[43] Khrushchev's secret speech, p. S49, discusses Ignatyev as minister. Yepishev, according to his *Bolshaya sovetskaya entsiklopediya* biography (3d ed.), became deputy MGB minister in 1951. His release as Odessa first secretary was reported in the 4 September 1951 *Pravda Ukrainy*.

[44] Khrushchev's secret speech, p. S45.

[45] Judging mainly by the appointments of deputy ministers reported in the *Sobraniye postanovleniy pravitelstva SSSR* and identifications of deputy ministers in the 1947 and 1950 election lists for the RSFSR and USSR Supreme Soviets.

[46] Khrushchev's secret speech, p. S46.

identify developments, it rarely illuminated police activity. Moreover, much of Beriya's activity probably involved intrigues behind the scenes with Stalin. It is probable that he played a key role in returning Malenkov to the leadership in 1948, as Khrushchev alleges, and in provoking the split with Tito, as Khrushchev charged; but again evidence is lacking.[47]

There is tenuous evidence that Beriya may have played a significant role in foreign policy, not only through his police apparat, but also through a highly placed CC official. His protégé V. G. Grigoryan, longtime editor of *Zarya vostoka*, the chief newspaper in Beriya's Georgia, began playing a leading role in foreign affairs in 1948–49, holding a high post either in the Cominform or the CC's foreign affairs organ—although his position has never been specifically identified in the press. As early as January 1948, Grigoryan, along with Yudin, represented the Soviet leadership at the congress of the Italian Communist party, according to the 7 January 1948 *Izvestiya*. In his memoirs, Ilya Ehrenburg mentioned that Grigoryan, who he says "held a rather high position," was the one who handled CC approval of Ehrenburg's speech to the April 1949 World Peace Congress.[48] At the 12 June 1950 Supreme Soviet session, Grigoryan was named a member of the foreign affairs commission of the Council of Nationalities—the only CC official on that body. He was identified as a "CC worker" in the 23 February 1950 *Pravda* and as head of an unnamed CC section in the 30 November 1950 *Vechernyaya Moskva*. In view of his foreign affairs activity, it appears that Grigoryan probably headed the CC's Foreign Policy Commission—indicating that Beriya had gained control over one of the key foreign policy posts in the wake of Zhdanov's fall.

The power of Zhdanov's other successful rival, Malenkov, lay in the party apparatus, especially in the Secretariat, which he rejoined in 1948. He quickly managed to persuade Stalin to bring his close protégé P. K. Ponomarenko, who had been demoted from Belorussian first secretary as a result of Zhdanov's machinations, to Moscow to join the Secretariat, thus changing the balance in that body.[49]

[47] On Beriya's role in Malenkov's return, see *Khrushchev Remembers*, pp. 252–53; on his role in the split with Tito, see Khrushchev's 26 May 1955 speech in Yugoslavia (*Pravda*, 27 May 1955).

[48] Ehrenburg, *Post-War Years 1945–1954*, vol. 6 of *Men, Years—Life* (London, 1966), p. 134.

[49] Ponomarenko, still identified as Belorussian premier at a 23–26 June 1948 Belorussian CC plenum (*Sovetskaya Belorussiya*, 27 June 1948), appeared with the

In early 1948 the Secretariat had consisted of Stalin, Zhdanov, and Zhdanov's allies Kuznetsov and Popov, plus the newly added Suslov. By July 1948 Zhdanov was inactive and his two allies were outnumbered by Malenkov, Ponomarenko, and Suslov. In February 1949 Kuznetsov was ousted and in late 1949 Popov also was dropped and replaced with Khrushchev. The result was a thoroughgoing change in the Secretariat.

Malenkov also used his restored power to reorganize and purge the cadres apparat taken from him during 1946. The Administration for Checking Party Organs (created in 1946) was combined with the Cadres Administration to form a Section for Party, Trade Union, and Komsomol Affairs. Although the reorganization, as usual, was not announced, Moscow *oblast* First Secretary G. M. Popov did refer to it in his report at the February 1949 party conference of Moscow *oblast,* declaring that "on Comrade Stalin's initiative," the CC "at the end of last year adopted a decree on reorganizing the party apparat," changing the CC cadres departments.[50] One of the deputy heads of the new section had already assumed his duties by the end of the year.[51]

Various leaders of the cadres and checking administrations brought in during Zhdanov's 1946 reorganization (A. N. Larionov, N. M. Pegov, G. A. Borkov) were transferred out by the end of the year. Larionov's 24 September 1960 *Pravda* obituary lists him as deputy chief of the Cadres Administration starting in September 1946 and first secretary of Ryazan starting in December 1948. Pegov, still identified as deputy chief of the Checking Administration in the 11 June 1948 *Pravda,* was identified in the 19 December 1948 *Pravda* as head of the Light Industry Section. Borkov, also still cited as deputy chief of the Checking Administration in the 11 June 1948 *Pravda,* was first secretary of Saratov *oblast* by early 1949. In

---

VKP Politburo and Secretariat members at the 25 July 1948 parade, indicating that he was now working in Moscow as part of the leadership. He was formally released as Belorussian premier on 12 September (*Sovetskaya Belorussiya,* 14 September 1948). The *Bolshaya sovetskaya entsiklopediya* (2d ed.) biography of Ponomarenko lists him as CC secretary, 1948–52.

[50] *Moskovskiy bolshevik,* 2 February 1949. The CPSU history issued in 1980, however, states that the reorganization of CC cadre organs, as well as of other CC sections, occurred already in July 1948 (*Istoriya kommunisticheskoy partii sovetskogo soyuza,* vol. 5, bk. 2, p. 220).

[51] The 5 March 1950 *Sovetskaya Belorussiya* biography of A. L. Dedov listed him as an official of the VKP CC's Cadres Administration starting in 1946, but as deputy head of the Section for Party, Trade Union, and Komsomol Affairs starting in 1948.

another reversal of the 1946 decisions, the journal *Partiynayha zhizn,* created in 1946, apparently stopped publication in mid-1948.[52]

In reorganizing the cadre organs, Malenkov also obtained a reversal of the principles of CC organization which had been adopted at the 1939 party congress on Zhdanov's initiative. At that time, Zhdanov, reporting on the party statute, complained that the CC's industrial-branch sections (*otdely*) were interfering too much with the work of economic organs and should be abolished. This measure was carried out and cadre work was concentrated in a single Cadres Administration.[53] In 1948, however, separate sections for industrial branches, such as light industry and heavy industry, were created,[54] and cadre work was thereafter shared by these sections. As Popov stated in explaining how the reorganization applied to the Moscow party organization, "the whole apparat," not just the cadres section, would now handle "selection, assignment and training of cadres."[55]

Also playing a key role in the post-Zhdanov leadership—though not as a ranking deputy to Stalin—was Suslov, the new CC secretary. Suslov rose fast in 1947, reorganized Zhdanov's ideological apparat during 1948, and played an important role in purging Zhdanov's faction in 1949.

Suslov had been only chief of the CC Bureau for Lithuania until March 1946, when he was transferred to work in the VKP CC and elected a member of the Orgburo (*Pravda,* 20 March 1946). Sometime during May or June 1947, Suslov became a CC secretary and by

[52] On 12 April 1954, *Pravda* announced resumption of publication of *Partiynaya zhizn.*

[53] See L. A. Maleyko's article " Iz istorii razvitiya apparata partiynykh organov" [From the history of the development of the apparatus of party organs] in *Voprosy istorii KPSS,* no. 2, 1976, pp. 111–22, for details on changes in the CC apparat in the 1920s and 1930s. After 1939, Malenkov pressed for a gradual return to branch organization in the CC apparat—for more on this, see Jonathan Harris, "The Origins of the Conflict Between Malenkov and Zhdanov: 1939–41," *Slavic Review,* June 1976, pp. 287–303, and Merle Fainsod, *How Russia Is Ruled* (Cambridge, Mass., 1964), pp. 197–200.

[54] *Pravda* on 19 December 1948 mentioned the Light Industry Section in identifying Pegov and on 20 March 1949 mentioned the Heavy Industry Section in reporting the death of one of the officials of the section. B. A. Abramov in the March 1979 *Voprosy istorii KPSS* (p. 64) writes that sections for heavy industry, light industry, machine building, transport, agriculture, and planning-finance-trade were among those set up in the 1948 reorganization.

[55] *Moskovskiy bolshevik,* 2 February 1949.

late 1947 was identified as Agitprop chief.[56] He was chosen to deliver the speech at the anniversary of Lenin's death on 21 January 1948—an honor performed by Aleksandrov the previous year. His first appearance at an ideological function was with CC secretaries Zhdanov, A. Kuznetsov, and Popov at the mid-January CC conference on music, held just before the CC adopted the important decree on Muradeli's opera "Great Friendship."[57] He gradually took over control of ideological affairs from Zhdanov, opening and speaking at a March 1948 conference of central press editors in Agitprop,[58] supervising a long conference of *oblast* newspaper editors in April,[59] and then directing a 19–20 May conference of leaders of the central publishing houses.[60] As already indicated, Suslov also took over supervision of the Cominform after the break with Yugoslavia.

As Zhdanov lost power in mid-1948, Suslov, acting in his capacity as CC secretary, thoroughly reorganized Agitprop and appointed new leaders. The Propaganda and Agitation Administration, still mentioned as late as 11 July 1948 in *Kultura i zhizn,* was changed into the Propaganda and Agitation Section in mid-July, its new title appearing first in the 21 July 1948 issue of *Kultura i zhizn.* Suslov turned over direct leadership of Agitprop to his deputy, D. T. Shepilov,[61] but continued to supervise it as CC secretary. The other leaders of the reorganized Agitprop included deputy heads L. F. Ilichev,[62] D. M. Popov (former Smolensk first secretary), L. A. Slepov (former *Pravda* editor for the department for party life), K. F. Kalashnikov (former head of the Agitation Section of Agitprop),

[56] The first indication of Suslov's elevation to CC secretary was when he joined members of the Politburo and Secretariat in the presidium of the RSFSR Supreme Soviet at its June session (*Moskovskiy bolshevik,* 21 June 1947). He had not been listed with other top leaders as a candidate for the RSFSR Supreme Soviet in February 1947 and was not among the leaders on the tribune at the May Day 1947 parade. He was first identified as CC secretary in the 24 September 1947 *Pravda* and first identified as Agitprop chief in the 23 November 1947 *Moskovskiy bolshevik.*

[57] *Partiynaya zhizn,* no. 6, March 1948. The 10 February 1948 decree is published in the 11 February 1948 issues of *Kultura i zhizn* and *Pravda.*

[58] *Kultura i zhizn,* 31 March 1948.

[59] *Kultura i zhizn,* 11 and 21 April 1948, and *Pravda,* 6 and 14 April 1948.

[60] *Kultura i zhizn,* 21 May 1948, and *Pravda,* 21 May 1948.

[61] Shepilov had been appointed deputy chief of Agitprop in late 1947, first being identified as such in the 21 November 1947 *Izvestiya.* He was still identified as deputy chief of the Propaganda Administration in the 21 May 1948 *Pravda* and was first identified as head of a CC section in the 31 July 1948 *Pravda.*

[62] Ilichev, still identified as chief editor of *Izvestiya* in the 4 December 1947 *Vechernyaya Moskva,* was first identified as deputy chief of Agitprop in the 24 March 1948 *Pravda* and retained this post after the reorganization.

and V. S. Kruzhkov (former director of the Marx-Engels-Lenin Institute). The organs of Agitprop also were affected. On 11 September 1948 *Kultura i zhizn* reported a CC decree declaring the work of Agit-prop's Academy of Social Sciences unsatisfactory and firing rector A. V. Mishchulin and deputy rector P. S. Cheremnykh. Mishchulin had been appointed rector by the 2 August 1946 decree that established the academy. The same issue of *Kultura i zhizn* reported a decree removing the chief editor of the satire magazine *Krokodil*.

The purge of Zhdanov's ideological organs expanded in 1949, with the replacement of the leaders of *Bolshevik, Pravda,* and *Kultura i zhizn.* The shake-up was triggered by a secret decree of 13 July 1949 which condemned Agitprop leader Shepilov for praising Voz-nesenskiy's book on the wartime economy of the USSR and ordered the removal of *Bolshevik* chief editor Fedoseyev and editorial board members Aleksandrov and Iovchuk for having promoted the book.[63] All three were dropped from *Bolshevik*'s editorial board in issue no 11, 1949 (signed to press 15 June 1949), along with five others, including deputy chief editor Kuzminov.[64] This purge reversed the late 1945 *Bolshevik* shake-up and cleared most of those associated with Zhdanov from the board.[65] Fedoseyev was reduced to a CC inspector.[66] Iovchuk was removed as Belorussian propaganda sec-retary in 1949, and Aleksandrov of course had already been demoted in 1947. The decree also resulted in Shepilov's removal as Agitprop head. When next mentioned in the press, he was only a CC inspec-tor.[67]

The *Bolshevik* shake-up was followed by replacement of the leaders of the other two main CC organs, *Pravda* and *Kultura i zhizn. Pravda* chief editor Pospelov, who had worked closely with Zhdanov, was demoted to director of the Marx-Engels-Lenin Insti-

[63] The existence of the decree was announced by Suslov in the 24 December 1952 *Pravda.*

[64] Kuzminov's obituary in the November 1979 *Voprosy ekonomiki* indicates that he was deputy chief editor until 1949.

[65] Only Ilichev, Kruzhkov, and Pospelov remained from the 1948 editorial board. S. M. Abalin became chief editor.

[66] Fedoseyev was identified as an inspector from 1950 to 1954 in the 1970 *Deputaty verkhovnogo soveta SSSR.*

[67] His last identifications as Agitprop head were in the 30 March 1949 issues of *Literaturnaya gazeta* and *Pravda* and in the 13 July 1949 CC decree. He was identified as a CC inspector in the 23 February 1950 *Pravda* and 29 November 1950 *Vecher-nyaya Moskva.*

tute in August 1949.[68] P. A. Satyukov, an apparent Zhdanov protégé from Gorkiy, was removed from the chief editorship of *Kultura i zhizn* and made secretary of *Pravda*'s editorial board,[69] and V. P. Stepanov, who has appeared to have ties with Suslov, became the new editor.[70]

During the 1949 purge, Suslov became the new chief editor of *Pravda*, in addition to being CC secretary, and he remained editor until late 1950 or early 1951.[71] Suslov thus became virtually the sole leader in the ideological field, especially since Agitprop appears to have remained without a director during this period. After Shepilov's 1949 ouster, no one was identified as Agitprop head until October 1952, when Komsomol First Secretary N. A. Mikhaylov assumed that position.[72]

Thus Suslov was a key beneficiary of the purge of Zhdanov's faction and clearly must have had a personal hand in purging the ideological apparat. However, unlike Beriya and Malenkov, who were publicly blamed for fabricating the Leningrad Case after their

[68] According to his biography in volume 34 of the *Bolshaya sovetskaya entsiklopediya,* 2d ed.

[69] For more on Satyukov's ties with Zhdanov, see Chapter 2. For Satyukov's transfer, see his biographies in the 1962 *Deputaty verkhovnogo soveta SSSR* and November 1971 *Zhurnalist.* Satyukov subsequently became a trusted Khrushchev aid as chief editor of *Pravda* from 1956 to 1964 and was fired upon Khrushchev's ouster because of his partisanship on behalf of Khrushchev (see Chapter 5).

[70] Stepanov's biography in the *Bolshaya sovetskaya entsiklopediya* yearbook for 1962 indicates that he worked in the CC apparat and was chief editor of *Gospolitizdat* after the war, became deputy editor of *Kultura i zhizn* sometime in 1949, and was made chief editor during the same year. In 1951 *Kultura i zhizn* ceased publication and Stepanov became deputy head of Agitprop (his biography lists him as editor until 1951, then as working in the CC apparat again; his first identification as deputy head of Agitprop was in the 28 April 1951 *Vechernyaya Moskva*). Stepanov rose to deputy chief editor of *Pravda* in 1961, became chief editor of *Kommunist* in 1962, and continued to be an active writer on ideological matters throughout the 1960s and 1970s. Stepanov edited the collections of Suslov's speeches published in 1972 and 1977, a job usually done by a leader's personal assistant.

[71] The 1971 *Bolshaya sovetskaya entsiklopediya* yearbook identified him as *Pravda* editor 1949–51; the *Bolshaya sovetskaya entsiklopediya* (3d ed.) entry on Suslov lists him as editor 1949–50. He was identified as *Pravda* chief editor as late as 21 November 1950 in *Moskovskaya pravda*, while Ilichev, who succeeded him as editor, was still listed as first deputy chief editor in the 1 December 1950 *Vechernyaya Moskva*. By 4 March 1951, when Khrushchev's "agro-town" proposal appeared in *Pravda*, Ilichev was chief editor (see Ilichev's speech in the 26 October 1961 *Pravda*).

[72] Mikhaylov was identified as Agitprop head 1952–53 in the *Bolshaya sovetskaya entsiklopediya* yearbooks for 1962 and 1966. Mikhaylov obviously assumed these duties when he was relieved as Komsomol first secretary and elected CC secretary in October 1952.

fall from power, Suslov has never fallen into disgrace and his role in these purges has never been publicly exposed. The most direct evidence of his involvement came only several years later, when Suslov in a 24 December 1952 *Pravda* article revealed the July 1949 CC decree and condemned Voznesenskiy and the Zhdanov clique. The December 1952 article was an attack on articles by Fedoseyev in the 12 and 21 December 1952 *Izvestiya*. Suslov recalled that Fedoseyev as editor of *Bolshevik* had approved praise of Voznesenskiy's book in *Bolshevik* and that this had been condemned as a "serious error" by the July 1949 decree. Fedoseyev quickly had to acknowledge the correctness of Suslov's charges in a 31 December 1952 letter printed in the 2 January 1953 *Pravda*. The reason for Suslov's belated public attack on Fedoseyev, Shepilov, and Voznesenskiy is not clear. Ironically, Suslov's attack appeared in *Pravda*, where Shepilov had just taken over as chief editor.

Thus, by early 1948 Zhdanov, the supposed spokesman for ideological orthodoxy, superpatriotism, and intolerance, had shown himself too "soft" to fit in with Stalin's turn toward dogmatism and the cold war. Stalin's growing dissatisfaction with Zhdanov had come to a head as a result of the Lysenko controversy and the split with Tito. In mid-1948, as Zhdanov was dying, Stalin's new favorites, Beriya, Malenkov, and Suslov, began erasing Zhdanov's influence and maneuvering their own people into positions of control. With Stalin's encouragement, they soon were to unleash a violent purge of moderates in science, who had enjoyed Zhdanov's protection, and then of Zhdanov's men in the party and government apparatus.

CHAPTER 4

Triumph of Reaction — 1949

With Zhdanov's death, the worst excesses of the postwar period began: Lysenko, with Stalin's endorsement, began his purge of science after the August 1948 session of VASKhNIL; on Stalin's orders, a vicious campaign against "cosmopolitans" began in January 1949; and, with Stalin's approval, Malenkov and Beriya organized the so-called Leningrad Case, which brought the arrest and eventual execution of Zhdanov's protégés Voznesenskiy and Kuznetsov, as well as other Leningrad leaders. Malenkov and Beriya became dominant in party and government politics, Suslov in ideology, Mitin and Yudin in philosophy, and Lysenko in biology, and with Stalin's encouragement, the harsh words of the Zhdanov era were replaced with harsh action. Apparently no one had been executed or even thrown into prison as a direct result of Zhdanov's campaign, but many executions occurred during the post-Zhdanov *Zhdanovshchina*. The death penalty, abolished in May 1947, during Zhdanov's heyday, was even publicly restored in 1950.

Whatever the roles of Malenkov, Beriya, and various dogmatists, however, the enthronement of chauvinism and the implementation of violent purges were first of all the work of Stalin himself. Moreover, Stalin's direct leadership of this type of vicious action increased after 1949 and, though Beriya may have put Stalin up to some of these actions in the late 1940s, by 1952 even Beriya could not be blamed for the new excesses since many of them were directed against Beriya himself.

Lysenko's Purge of Science

The first of the post-Zhdanov crackdowns actually began shortly before Zhdanov died—at the infamous August 1948 session of VASKhNIL. Armed with Stalin's endorsement of his position as the only one permissible for loyal Soviet scientists, Lysenko forced biologists to take sides for or against him and then revealed that those taking the latter course were also disagreeing with Stalin and the party. There followed a massive purge in biology and other branches of science to remove from influence anyone who had opposed Lysenko and "Russian" science.

Although Lysenko had long been president of VASKhNIL, the few Lysenko supporters among the academy's seventeen members (such as V. P. Mosolov, I. G. Eykhfeld, I. V. Yakushkin) were outnumbered by his outspoken foes (such as P. M. Zhukovskiy, I. I. Shmalgauzen, and B. M. Zavadovskiy). Since Lysenko was unable to obtain the election of his supporters to VASKhNIL, he avoided elections. However, on 15 July 1948 Stalin resolved the imbalance when he simply appointed thirty-five new members from a list compiled by Lysenko which included mainly his supporters (I. I. Prezent, M. A. Olshanskiy, D. A. Dolgushin, P. A. Vlasyuk, B. P. Bushinskiy, S. F. Demidov, I. D. Kolesnik, I. D. Laptev, P. P. Lukyanenko).[1]

Lysenko used the session (31 July-7 August) to provoke a debate between his supporters and his opponents. He opened the session with a long report in which he rejected the chromosome theory of heredity and argued that there were two "irreconcilable" biologies—a Western bourgeois, "Weismannist-Mendelist-Morganist" biology (upholding the chromosome theory) and a "Michurinist, Soviet" biology (upholding inheritance of acquired characteristics). Lysenko attacked his critics by name (I. I. Shmalgauzen, B. M. Zavadovskiy, P. M. Zhukovskiy, A. R. Zhebrak, N. P. Dubinin, etc.), and his supporters (members and nonmembers of VASKhNIL) did the same. Several geneticists (Zavadovskiy, Zhukovskiy, Shmalgauzen, I. A. Rapoport, S. I. Alikhanyan, V. S. Nemchinov) tried cautiously to defend the chromosome theory while

---

[1] *Pravda*, 28 July 1948; also see Medvedev's *The Rise and Fall of T. D. Lysenko*, pp. 114–16.

denying that they were geneticists and arguing that there was a "third line"—neither Western nor Lysenkoist, but Soviet.

On the last day of the session, however, before beginning his concluding speech, Lysenko announced that the CC had approved his opening report even before the VASKhNIL meeting had opened (*Pravda*, 8 August 1948)—that is, confirmed that biologists had to be either Lysenkoist or "Mendelist-Morganist." Thereafter, most of his foes conceded the accuracy of Lysenko's thesis that there were indeed two sciences in the world and therefore recognized the need to pledge allegiance to Soviet science. Zhukovskiy, Alikhanyan, and I. M. Polyakov renounced their Mendelist-Morganist views at the session and promised to fight for Michurinism (*Pravda*, 8 August 1948), while Zhebrak renounced his anti-Lysenko position in a 9 August letter printed in the 15 August 1948 *Pravda*. A few—notably V. S. Nemchinov—defiantly refused to change their stands (as pointed out in *Pravda*, 8 August).

A session of the Academy of Sciences Presidium was immediately convened (24–26 August) to follow VASKhNIL's lead in condemning genetics, recognizing the dominance of Lysenkoism in biology, and acknowledging the existence of two hostile sciences. L. A. Orbeli, the academic secretary of the biology division, delivered a keynote speech in which he admitted error in not backing Lysenko and, following harsh attacks on him by Lysenko supporters, he offered to resign as head of the biology division.[2] At the VASKhNIL session, Orbeli had refused to speak after hearing Lysenko's speech.[3] The academy presidium session concluded by issuing a decree in which it criticized itself for supporting the "reactionary Weismannist school" rather than the Michurinists. The decree also censured Orbeli and ordered a number of organizational actions: (1) Orbeli was removed as head of the biology division and replaced with A. I. Oparin; (2) Lysenko was added to the bureau of the biology division; (3) Shmalgauzen was removed as director of the Institute of Evolutionary Morphology; (4) Dubinin's cytogenetics lab in the Institute of Cytology, Histology, and Embriology was dissolved; and (5) Weismannists-Morganists were ordered purged from scientific councils of biology institutes and editorial collegia of biology journals.[4]

[2] *Vestnik akademii nauk SSSR*, no. 9, 1948, p. 170.
[3] As he acknowledged in his concluding speech, ibid., p. 167.
[4] Ibid., pp. 21–24.

Immediately after the VASKhNIL meeting, according to Zhores Medvedev, hundreds of anti-Lysenko scientists were fired or demoted on a variety of charges.[5] Labs were closed and "nearly every scientist" had to appear and declare his views before special commissions created to seek out anti-Lysenkoists.[6] Medvedev reports that as early as 23 August 1948, an order from S. V. Kaftanov, the minister of higher education, fired professors Shmalgauzen, Zavadovskiy, D. Sabinin, dean of the faculty S. D. Yudintsev, associate professor Alikhanyan, and others at Moscow State University, as well as Golubev, Zhebrak, and others from the Timiryazev Agricultural Academy.[7] In an article in the 8 September 1948 *Izvestiya*, Kaftanov declared that his ministry had been too lax on anti-Michurinists and singled out for attack scientists at Leningrad State University, the Timiryazev academy, and elsewhere. He announced that Nemchinov had been fired as director of the Timiryazev academy and replaced with V. N. Stoletov, that Lysenko himself had taken over the academy's department of genetics and selection from Zhebrak, that Prezent had replaced the anti-Lysenkoist Yudintsev as dean of the biology faculty at Moscow State University, and that M. Ye. Lobashev had been removed as dean of the biology faculty at Leningrad State University. In his 31 July 1948 report, Lysenko had complained that Mendelism-Morganism dominated in educational institutions, had ridiculed Dubinin's genetics experiments with fruit flies, and had called on the Ministry of Agriculture to do something about Zhebrak's genetics research. After August 1948, according to Medvedev, textbooks were destroyed, the Ministry of Agriculture closed down all genetics research in agriculture, and even the fruit flies used in genetics research were wiped out.[8] Dubinin, a special target of Lysenko,[9] had not attended the August 1948 session of VASKhNIL, for the session had been prepared in secret and Dubinin's invitation had arrived only after he had left on vacation.[10] When he returned, he was out of work.

Persecution continued for some time and many were banned from

[5] Medvedev, *The Rise and Fall of T. D. Lysenko*, p. 103.
[6] Ibid., p. 124.
[7] Ibid., pp. 124–25.
[8] Ibid., pp. 125–26.
[9] Lysenko had protested when Dubinin was elected a corresponding member of the Academy of Sciences in 1946, according to Lev Kokin's article in *Moskva*, no. 10, 1965.
[10] According to V. Gubarev in the 27 February 1965 *Komsomolskaya pravda*.

meaningful scientific work for years. Lysenko's ally Prezent attempted to get Lobashev and other anti-Lysenkoists expelled from the party, in addition to their dismissal from Leningrad State University.[11] Sabinin, continually persecuted after 1948, finally committed suicide in 1951.[12]

Though dismissals were widespread, actual arrests were few, as both Medvedev and David Joravsky, another close student of Soviet science, indicate in their detailed studies of Lysenko.[13] Apparently only two prominent anti-Lysenko scientists were arrested: D. D. Romashev, head of a fish-genetics lab, and V. D. Efroimson, docent at Kharkov University—both of whom were released in 1955.[14] Nevertheless, even if few were imprisoned and no one killed in this purge, the damage it inflicted on Soviet science was devastating and the 1948–49 purge of geneticists became a landmark in Stalinist repression just as the *Zhdanovshchina* had.

The Anticosmopolitanism Campaign

The next big campaign was the attack on cosmopolitans which began in January 1949. In practice, the term "cosmopolitan" usually applied to Jews, although it could refer to anyone who was "internationalist" or "unpatriotic," judged by the superpatriotic standards of the day. As was later revealed, this campaign was launched on Stalin's personal initiative.

Although one of Zhdanov's themes had been to exalt things Russian and attack those who bowed to Western influence, the term "cosmopolitan" came into extensive use only as Zhdanov faded. A few attacks on Zhdanov's protégés as "cosmopolitan" and "antipatriotic" had already appeared in July 1948, but the press attacks on "rootless cosmopolitans" really began only in January 1949.[15]

[11] Medvedev, *The Rise and Fall of T. D. Lysenko*, p. 267.

[12] Ibid., p. 128. Harassment of Orbeli, Lobashev, and V. V. Sakharov after 1948 is described by émigré scientist Raisa Berg in "Povest o genetike" [A story about genetics], *Vremya i my* (Tel Aviv), no. 50, 1980, pp. 162–202.

[13] Medvedev, *The Rise and Fall of T. D. Lysenko*, pp. 128–29, and David Joravsky, *The Lysenko Affair* (Cambridge, Mass., 1970), pp. 118, 388.

[14] Medvedev, *The Rise and Fall of T. D. Lysenko*, pp. 128–29, and Joravsky, *The Lysenko Affair*, pp. 321, 325. Romashev was sharply attacked by Stoletov at the August 1948 session as Dubinin's "closest coworker."

[15] Arrests of some leading Jews had occurred earlier in 1948, but this appeared separate from the later campaign. According to Khrushchev, the arrests were part of Stalin's reaction to a proposal by the Jewish Anti-Fascist Committee to create a Jewish Autonomous Republic in the Crimea (see note 36, Chapter 3).

This campaign was launched after certain writers made unpublished attacks on Jewish theater critics as antipatriotic cosmopolitans in late 1948. According to Konstantin Simonov, in 1948–49 the deputy general secretary of the Writers' Union and chief editor of *Novyy mir,* critics who had pointed out real weaknesses in drama were assailed by their foes at a late 1948 Writers' Union board plenum for criticizing plays from hostile positions. "Unconscionable demagogic methods" were used to "create the appearance of a 'hostile' position" by these critics and to construe their criticism as applying to all aspects of Soviet society. They were accused of bowing to the West and being cosmopolitan and antipatriotic.[16]

According to Simonov, the accusations were then picked up and officially approved by an editorial article in the 28 January 1949 *Pravda.* Simonov recalls that "the views and evaluations expressed in this article, the initiator of which, as was quite well known in literary circles, was I. V. Stalin directly, led to consequences very serious for literature."[17] The 28 January 1949 *Pravda* article and a similar editorial article in the 30 January 1949 *Kultura i zhizn* denounced "rootless cosmopolitans" among theater critics.[18] The *Pravda* editorial article "About an antipatriotic group of theater critics" and the *Kultura i zhizn* editorial article "On hostile positions" criticized such "rootless cosmopolitans" as Yu. Yuzovskiy, A. Gurvich, Ye. Kholodov (Meyerovich), and A. Borshchagovskiy—all of whom clearly were Jewish. The articles revealed that at a meeting of

[16] These events were recalled by Simonov in the December 1956 *Novyy mir.* He is apparently referring to the mid-December 1948 plenum of the Writers' Union board, which heard Writers' Union Secretary A. Sofronov speak on improvement of theater repertoires. However, the published accounts of Sofronov's 17 December report (*Literaturnaya gazeta,* 22 December 1948) and of the 17–18 December discussion (*Literaturnaya gazeta,* 25 December 1948) contain virtually no references to cosmopolitanism.

[17] *Novyy mir,* December 1956, p. 250. Ilya Ehrenburg also attributes the campaign directly to Stalin. He later wrote: A. A. Fadeyev (then the general secretary of the Writers' Union) "told me that the press campaign against the 'unpatriotic critics' had been launched on Stalin's instructions. But a few weeks later he summoned the editors and said: 'Comrades, the divulging of literary pseudonyms is inadmissible, it smells of anti-Semitism.' Rumour attributed the arbitrary measures to those who carried them out while Stalin was thought to have been a restraining influence. By the end of March he apparently decided that the job was done" (Ehrenburg, *Post-War Years 1945 – 1954,* p. 133).

[18] A preview of the attacks appeared in a *Kultura i zhizn* editorial (11 January 1949), which attacked the writings of N. Melnikov and used the device of giving his real name—Melman—in parentheses to indicate that he was Jewish rather than Russian.

dramatists in late November 1948, some of these critics had wrongly criticized certain plays and that the press—including *Pravda, Izvestiya, Kultura i zhizn, Literaturnaya gazeta,* and *Komsomolskaya pravda*—had promoted these cosmopolitan views by publishing their articles. On 11 February 1949, *Pravda* and *Kultura i zhizn* reported that a meeting of the Writers' Union party organization had been called for 9–10 February to discuss the *Pravda* and *Kultura i zhizn* attacks and had denounced the theater critics. In the following weeks, *Pravda* and *Kultura i zhizn* carried numerous articles denouncing cosmopolitans in the theater, music, architecture, poetry, cinema, and philosophy. Simonov himself wrote a *Pravda* article (28 February 1949) denouncing cosmopolitans and later admitted that he and other leaders of the Writers' Union, lacking the courage to resist, had echoed the January *Pravda* article and added their own "grossly unjust" accusations.[19]

The anticosmopolitanism campaign reinforced the campaign to condemn Zhdanov's clique of philosophers as unpatriotic. The 2 March 1949 *Literaturnaya gazeta* reported a two-day meeting of the Institute of Philosophy's scholarly council which denounced the "anti-Marxist" and "cosmopolitan" errors of Kedrov and asked that the Academy of Sciences remove him from the institute. G. F. Aleksandrov, director of the Institute of Philosophy, delivered the main report on Kedrov's "mistaken views" in philosophy and science, accusing him of using *Voprosy filosofii* to support Lysenko's opponents.[20] Maksimov called Kedrov a "mouthpiece for reactionary ideas," and Mitin attacked Kedrov as "nihilistic" toward Russian culture and philosophy.

In a *Literaturnaya gazeta* article of 9 March, Zhdanov's foe Mitin seized upon the anticosmopolitan campaign to demand that philosophy also be cleansed of such "rootless cosmopolitans" as B. Kedrov, Z. Kamenskiy, M. Selektov, I. Kryvelev, V. Goffenshefer, and Ya. Chernyak. Mitin concentrated on his rival Kedrov, whom he labeled the "ideological inspirer" of cosmopolitans in philosophy. He argued that Kedrov was a cosmopolitan because he opposes the

---

[19] *Novyy mir,* December 1956, pp. 249–50. One of the Jews under attack, A. Borshchagovskiy, was chief of *Novyy mir*'s department of the literature and art of the peoples of the USSR, working under Simonov (his appointment had been announced in the 20 November 1946 *Kultura i zhizn*).

[20] Aleksandrov did not get off completely free either: the account notes that speakers criticized the Institute of Philosophy leadership's "liberal attitudes" toward Kedrov's views.

struggle against "cringing before everything foreign" and "propagandizes the fictitious priority of foreigners," because he has a "nihilistic attitude toward Russian national culture" and "ignores the historic role of great Russian scientists," and because he has promoted the concept of a "world science that knows no national or state borders."

An editorial article in the 10 March 1949 *Kultura i zhizn,* discussing the recent Institute of Philosophy session on cosmopolitanism in philosophy, attacked the "cosmopolitan views" of B. Kedrov, M. Rozental, I. Kryvelev, M. Selektov, and Z. Kamenskiy. Focusing mainly on Kedrov, the article argued that he had downgraded the contribution of Russian scholars and overrated Western philosophy and science. It complained that even after the decisions in the 1947 philosophy debate (apparently meaning the condemnation of Aleksandrov for neglecting the role of Russian philosophy), Kedrov had ignored "questions of the priority of Russian science" and argued that there is only one "world science" and that to view each country's philosophy as a national philosophy is un-Marxist. The article stated that Kedrov had asserted that before the 1840s there had been no philosophy except West European philosophy and that anyone studying the history of Russian philosophy was therefore wasting his time. The same piece accused Rozental of having provided a "tribune" for the "hostile views" of the cosmopolitan literary critics Yuzovskiy, Levin, Altman, and Gurvich while he had served as editor of the journal *Literaturnyy kritik.* It was alleged that the journal "in every way ran down Russian literature and Russian theater," and that the cosmopolitan literary critics exalted Rozental "as their ideological leader and 'theoretician.'" In addition, it was argued that Kryvelev, until his recent dismissal as secretary of *Voprosy filosofii*'s editorial board, had had *Voprosy filosofii* print "cosmopolitan" articles by Kamenskiy and non-Lysenkoist articles by biologist I. Shmalgauzen, "one of the ringleaders of Weismannism-Morganism in the USSR." Kryvelev allegedly "attacked those who spoke about the leading role of the great Russian people in our society, accusing the latter of nationalism."[21]

Kedrov responded by writing a letter published in the 22 March 1949 *Kultura i zhizn,* in which he acknowledged that the paper's 10 March attacks on him, both the accusation of neglecting the priority

[21] Judging by his first name and patronymic, Iosif Aronovich, Kryvelev is Jewish. There is no evidence that Kedrov is Jewish, however.

of Russian science and the specific charge of ignoring Lysenko, were correct. He admitted that he had also erred in failing to mention the significance of the August 1948 VASKhNIL session in his October 1948 brochure on natural science.

When the next issue of *Voprosy filosofii* appeared (in June 1949[22]), chief editor Kedrov had been reduced to just a member of the editorial board and board secretary Kryvelev had been dropped altogether. The new chief editor was D. I. Chesnokov, who had assailed the Aleksandrov–Fedoseyev clique in the June 1947 philosophy debate. Also added to the board were Lysenko allies M. B. Mitin and V. N. Stoletov. This issue carried a sharp editorial attack on Kedrov, Kryvelev, and Kamenskiy as cosmopolitans, an article by Maksimov assailing Markov and Kedrov, and also articles by G. F. Aleksandrov and M. T. Iovchuk denouncing cosmopolitans and condemning their former colleague Kedrov for defending cosmopolitan views.

The Leningrad Case

The most far-reaching and bloody episode following Zhdanov's fall was the so-called Leningrad Case, in which Zhdanov's leading protégés, Politburo member N. A. Voznesenskiy and CC Secretary A. A. Kuznetsov, as well as some other Zhdanov men, were not only ousted from their positions but executed. Zhdanov's rivals Beriya and Malenkov organized the case with Stalin's connivance; it started in February 1949, when the first dismissals were carried out, and ended in September–October 1950, when the main victims were shot.

The charges against the Leningraders were not made public; indeed, the existence of the "case" itself was not revealed publicly until after Stalin's death. The charges were kept so secret that even Khrushchev, a member of the Politburo, appeared to have only a vague knowledge of them. In his recollections Khrushchev states that he himself never saw the indictments in the Leningrad Case but only concluded from remarks by Beriya and Malenkov that the charges were Russian nationalism and opposition to the CC.[23]

[22] *Voprosy filosofii*, no. 2, 1948, was signed to press on 27 October 1948 under Kedrov. The next issue, labeled no. 3, 1948, was not signed to press until 1 June 1949, under Chesnokov.

[23] *Khrushchev Remembers*, p. 256. The tapes on which Khrushchev's recollections

Initially, stories were spread that Leningrad First Secretary P. S. Popkov and others in Leningrad were guilty of corruption and had been protected by CC Secretary Kuznetsov, the former Leningrad first secretary.[24] In 1956 party leaders told C. L. Sulzberger that Voznesenskiy and Kuznetsov had been shot for attempting to create a separate party organization for the RSFSR centered in Leningrad instead of Moscow.[25]

Whatever the exact nature of the charges, however, they surely involved treason. The death penalty had been abolished by a 26 May 1947 ukase, but it was restored "by way of exception" for "spies" and "traitors" on 12 January 1950 (*Pravda*, 13 January 1950). Subsequently, in September 1950, a military court sentenced Voznesenskiy, Kuznetsov, and others to death, presumably on the basis of the restored death penalty.[26]

The charge of treason may have stemmed from the contacts that the Leningraders had had with a delegation of Yugoslavs led by Milovan Djilas, which visited the city in January 1948 and received very friendly treatment. The Leningrad leaders apparently provided their visitors with some information that they had not received during their stay in Moscow, for Stalin's 4 May 1948 letter to the Yugoslavs noted that Djilas got information about party and state

---

were based include a more detailed version of Khrushchev's discussion of the Leningrad Case than is in *Khrushchev Remembers,* although the remarks are disjointed and unclear. Discussing the intrigues of Malenkov and Beriya against Kuznetsov, Voznesenskiy, and Kosygin, Khrushchev mentions that Malenkov and Beriya apparently prepared documents aimed at undermining Stalin's trust in these three. The documents, he says, accused them of "nationalism, supposedly Russian nationalism," and "setting themselves against the CC." After some digressions, he again discusses some documents—apparently these same documents—which he was supposed to sign and turn over to Stalin. He relates vaguely that before signing he first consulted Malenkov. "Well, you know, he told me, the basic accusation against the Leningraders, that is, it was alleged that they, so to say, have been showing some sort of independent initiative, that is, that they set up some sort of fair in Leningrad, that is, they organized it. And there they sold off goods, hard-to-sell goods, as one always sells at fairs.... I didn't see any political crime here, any manifestation of Russian (*rossiyskiy*) nationalism in this, so to speak, or Russian (*russkiy*) nationalism. Besides, we had done the same thing in Kiev. We sold, that is, also sold [goods].... That was the kind of fair it was, we sold off unsold goods that no one had bought in the stores, that is" (Russian-language transcripts, pt. I, pp. 985–89).

[24] According to Boris Nicolaevsky in "The Abakumov Case," *New Leader,* 10 January 1955.

[25] C. L. Sulzberger, *The Big Thaw* (New York, 1956), pp. 47–48.

[26] According to F. Leonov in a 20 February 1965 *Pravda* article, Kuznetsov and others were accused of "treason against the motherland," and in September 1950 a military collegium sentenced Kuznetsov, Voznesenskiy, and others to be shot.

work from local Leningrad organs during his visit. He went on to point out that when the Soviets gathered such information in Yugoslavia, the Yugoslavs called it spying.[27] Although Stalin did not object to Djilas's information gathering in the May 1948 letter, it would be easy, once the Yugoslavs were identified as enemies of the USSR, to construe the provision of such information as treason. Describing the visit in his book, Djilas praises the Leningrad leaders, declaring that he got along well with them and indicating that it was a relief to visit them—quite in contrast to the coldness officials in Moscow were already displaying toward the Yugoslav delegation.[28]

Beriya and his protégé V. S. Abakumov, the minister of state security, took the leading roles in fabricating the Leningrad Case, though Malenkov played an active role, as well. After Beriya's mid-1953 ouster, the blame for the case initially centered on Beriya and Abakumov. Abakumov was tried and executed for this and other crimes in December 1954. According to the 24 December 1954 *Pravda* account of his 14–19 December trial (held in Leningrad), Abakumov had been placed in his job by Beriya and, carrying out Beriya's orders, had "fabricated the so-called 'Leningrad Case'" and other cases by forcing confessions of "grave state crimes" from arrested officials. Khrushchev's 1956 secret speech blamed Beriya and Abakumov for fabricating the evidence at Stalin's behest.[29]

As Malenkov's power decreased, however, he too began to receive blame for the case, at first secretly, then publicly. At the January 1955 CC plenum, during Khrushchev's attack on Malenkov, Malenkov's involvement in the case was brought up, although this information was not then made public. Giuseppe Boffa, who apparently saw transcripts of the speeches that were delivered at the January 1955 plenum, states that they indicated that Malenkov had admitted "moral coresponsibility" for the Leningrad Case.[30] Weakened by these revelations, Malenkov soon resigned as premier. Once Malenkov had been expelled from the Politburo in the "Anti-Party

[27] *The Soviet-Yugoslav Dispute,* p. 41. The visit is discussed in Vladimir Dedijer's *Tito Speaks,* pp. 321–22.

[28] Djilas, *Conversations with Stalin,* pp. 168–69.

[29] Khrushchev's secret speech, p. S46.

[30] Giuseppe Boffa, *Inside the Khrushchev Era* (New York, 1959), pp. 26–27. Translated from *La grande svolta.* Boffa, a correspondent in Moscow, also says that in 1957 it was revealed that Kuznetsov had actually been arrested in Malenkov's office. Seweryn Bialer, then an official of the Polish party, also saw party documents indicating that Malenkov was blamed for the Leningrad Case at the January 1955 plenum ("I Chose Truth," *News from Behind the Iron Curtain,* October 1956, p. 7).

Group" purge of mid-1957, Khrushchev and others began to blame him publicly for the Leningrad Case. In a Leningrad speech reported in the 7 July 1957 *Pravda*, Khrushchev declared that Malenkov had been "one of the main organizers" of the Leningrad Case. At the 2 July 1957 meeting of the Leningrad *aktiv*, held to denounce the "Anti-Party Group," I. N. Turko, former second secretary of Leningrad *oblast*, revealed that in February 1949 Malenkov had tried to intimidate him into signing forged documents that related to the Leningrad Case.[31] At the 1961 party congress, Leningrad First Secretary I. V. Spiridonov, Gorkiy First Secretary L. N. Yefremov, and CC Secretary A. N. Shelepin blamed Malenkov for the Leningrad Case (*Pravda*, 20, 22, 27 October 1961). Official versions now portray Beriya as primarily responsible, with Malenkov in a supporting role. For example, the entry on Voznesenskiy in the *Sovetskaya istoricheskaya entsiklopediya* declares that the "so-called Leningrad Case" was "fabricated by L. Beriya with the help of G. Malenkov and with Stalin's sanction."[32] Khrushchev also treats Beriya as the main culprit and Malenkov as his accomplice.[33]

The motivation of Beriya and Malenkov in concocting the case was clearly to destroy the rival Zhdanov faction. In his recollections, Khrushchev contends that Beriya and Malenkov regarded Kuznetsov and Voznesenskiy as their rivals and, fearing that Stalin would replace them with these two newcomers, did everything they could to turn Stalin against the two men.[34] Khrushchev had already suggested this motivation in his 1956 secret speech, although at that time he mentioned only Beriya, for Malenkov was still a Politburo

[31] On 27 August 1946 (*Pravda*, 28 August 1946), Turko had been transferred to first secretary of Yaroslavl as part of Zhdanov's moves against Malenkov. (Yaroslavl First Secretary Larionov was then relieved and brought to Moscow to become deputy chief of the reorganized CC Cadres Administration after Malenkov's dismissal as CC secretary in charge of cadres.) In February 1949, Turko was summoned from Yaroslavl to meet with Malenkov, who "tried to force me to sign a conscious falsification" (*Leningradskaya pravda*, 5 July 1957). The CC took action against Turko, issuing a decree on shortcomings in Yaroslavl's party educational work. (The *Istoriya kommunisticheskoy partii sovetskogo soyuza*, vol. 5, bk. 2, p. 631, gives the date of this decree as 24 January 1949.) On 8 February 1949, *Pravda* reported that a recent plenum of the *oblast* party committee had heard Turko report on this decree and admit the *oblast*'s shortcomings. Turko was soon fired and replaced with G. S. Sitnikov, a former colleague of new Leningrad First Secretary V. M. Andrianov (he had served as executive committee chairman of Sverdlovsk *oblast* while Andrianov was *oblast* first secretary).

[32] Vol. 3, published 1963.

[33] *Khrushchev Remembers*, pp. 251–52.

[34] Ibid., pp. 250–52, 256.

member in more or less good standing. He explained that the rise of Voznesenskiy and Kuznetsov had alarmed Beriya, who then began intriguing to discredit them, supplying Stalin with "slanderous material" against them.[35] Khrushchev revealed that Stalin had placed CC Secretary Kuznetsov in charge of state security, thereby making him a rival to Beriya in a sphere that had formerly belonged to Beriya alone.

The main victims of the Leningrad Case were Politburo member Voznesenskiy, CC Secretary Kuznetsov, and Leningrad First Secretary Popkov, although RSFSR Premier M. I. Rodionov and Leningrad city Second Secretary Ya. F. Kapustin may also have been put on trial with them. In the 20 February 1965 *Pravda*, F. Leonov stated that in September 1950 Kuznetsov, Voznesenskiy, and others were sentenced by a military court to be shot. Official biographies give the dates of death as 30 September 1950 for Voznesenskiy,[36] 1 October 1950 for Kuznetsov,[37] and 1 October 1950 for Popkov[38]—dates suggesting that all were sentenced at the same time. In his secret speech, Khrushchev listed Voznesenskiy, Kuznetsov, Popkov, and Rodionov as dying in the Leningrad Case.[39] The Leningrad *Entsiklopedicheskiy spravochnik,* issued in 1957, suggests (pp. 138–39) that Kapustin was also one of those tried, since it declares that in the Leningrad Case, Voznesenskiy, Kuznetsov, Kapustin, Popkov, and others were "slandered and condemned." There are no official biographies or rehabilitation articles that supply the dates of death for Rodionov and Kapustin, so it is unclear whether they died at the same time as the others, though they were removed from their posts with the others in late February or early March 1949. The fact that Kapustin was the only Leningrad city or *oblast* secretary removed at about the same time as Popkov emphasizes his special involvement in

---

[35] Khrushchev's secret speech, pp. S45–46.

[36] This is the date given in his biographies in the *Bolshaya sovetskaya entsiklopediya* (3d ed.), *Malaya sovetskaya entsiklopediya, Sovetskaya istoricheskaya entsiklopediya, Ukrainska radyanska entsiklopediya,* and *Ekonomicheskaya entsiklopediya* (vol. 1, 1972).

[37] Official biographies of Kuznetsov long persisted in listing his date of death as 1949: in the *Bolshaya sovetskaya entsiklopediya* (2d ed., vol. 51, printed 1958), *Malaya sovetskaya entsiklopediya* (1959), *Ukrainska radyanska entsiklopediya* (1962), and even the *Sovetskaya istoricheskaya entsiklopediya* (printed as late as 1965). However, the most recent versions (*Bolshaya sovetskaya entsiklopediya,* 3d ed., 1973) list it as 1 October 1950.

[38] *Bolshaya sovetskaya entsiklopediya,* 3d ed., published 1975.

[39] Khrushchev's secret speech, p. S45.

the case.[40] All of these men suffered because of their former ties to Zhdanov, although in Voznesenskiy's case there were additional reasons and he will be discussed separately below.

Kuznetsov was an especially close protégé of Zhdanov,[41] serving as Zhdanov's top deputy in Leningrad for several years (second secretary of Leningrad city from February 1938 to January 1945 and second secretary of Leningrad *oblast* from about 1943 to January 1945, while Zhdanov was first secretary of both city and *oblast*) and succeeding Zhdanov as *oblast* and city first secretary in January 1945. In March 1946 Kuznetsov followed Zhdanov to Moscow, becoming a CC secretary. According to Khrushchev, he was entrusted with supervision of state security, thereby becoming a threat to Beriya. Kuznetsov was removed as CC secretary and Orgburo member in February 1949,[42] and, according to two sources, he then became secretary of the CC's Bureau for the Far East.[43]

Popkov was also close to Zhdanov and Kuznetsov, with whom he served as city executive committee chairman from 1939 until March 1946, at which time he himself became Leningrad city and *oblast* first secretary. His friendliness to the Yugoslav delegation in 1948 may have been used against him, and at some point in the second half of February 1949 he was removed and disappeared.[44]

Rodionov was not a Leningrader but a native of Gorkiy *oblast,* where he rose under longtime *oblast* First Secretary Zhdanov and eventually served as *oblast* first secretary from 1940 to June 1946.[45] His promotion to RSFSR premier in June 1946 (he succeeded Zhdanov protégé Kosygin) presumably was carried out on

---

[40] Popkov was removed at a February plenum (*Leningradskaya pravda,* 24 February 1949) and Kapustin at a 4 April Leningrad city plenum (*Leningradskaya pravda,* 5 April 1949). The other *oblast* and city secretaries remained in their posts for several more months.

[41] In the 14 November 1947 *Pravda,* he was described as Zhdanov's "closest comrade-in-arms" during the war.

[42] According to volume 51 of the *Bolshaya sovetskaya entsiklopediya* (2d ed.) and to F. Leonov in the 20 February 1965 *Pravda.*

[43] *Ukrainska radyanska entsiklopediya* and Leningrad's *Entsiklopedicheskiy spravochnik* (issued 1957).

[44] The 24 February 1949 *Leningradskaya pravda* briefly reported that a joint plenum of the Leningrad city and *oblast* committees "in recent days" had removed Popkov and elected V. M. Andrianov to replace him as first secretary. Popkov's last public appearance as Leningrad first secretary was on 31 January 1949 (*Leningradskaya pravda,* 1 February 1949).

[45] Biographic details on Rodionov are given in a 2 November 1967 *Sovetskaya Rossiya* article marking his sixtieth birthday.

Zhdanov's recommendation. Indeed, Zhdanov presided at the RSFSR Supreme Soviet session which elected him (*Pravda,* 26 June 1946). Rodionov was removed with Voznesenskiy and Kuznetsov in March 1949,[46] and according to Khrushchev's secret speech, he perished along with them. The Gorkiy party organization was also purged in 1949 and 1950, following a 23 March 1949 CC decree that condemned the leaders of Gorkiy *oblast* for "big mistakes" and ordered a shake-up.[47] Gorkiy First Secretary S. Ya. Kireyev was soon removed, and in the 6 May 1950 *Pravda* the new first secretary, D. G. Smirnov, attacked the "former leading officials" of Gorkiy for a variety of misdeeds. The Gorkiy decree attacked "former" RSFSR Premier Rodionov for following "the harmful practice of acting as guardian over Gorkiy *oblast* and by his incorrect actions teaching some leaders of party and soviet organs to be dependents."[48] Gorkiy First Secretary L. N. Yefremov later stated, at the 1961 CPSU congress, that during "the sad period of the provocatory 'Leningrad Case,' fabricated by Malenkov," a number of Gorkiy leaders also suffered persecution and "groundless political accusations" were made against leading party and soviet organs in Gorkiy.[49] Khrushchev notes that those arrested in Moscow in connection with the Leningrad Case were mainly party officials "promoted by Zhdanov from the Gorkiy organization to the Moscow organization."[50]

One of the intended victims of the Leningrad Case—Politburo member and Deputy Premier Kosygin—escaped. Khrushchev later declared that since Kosygin was related to Kuznetsov and since other victims had accused him of involvement in the Leningrad plot, "his life was hanging by a thread."[51] Kosygin had risen in Leningrad under Zhdanov, eventually becoming chairman of the city executive com-

[46] *Pravda* on 11 March announced that the Presidium of the RSFSR Supreme Soviet had removed Rodionov as premier. The date was not specified.
[47] The decree was published in the *KPSS v rezolyutsiyakh i resheniyakh syezdov, konferentsiy i plenumov TsK,* vol. 6, 1971, pp. 277–80.
[48] Ibid., p. 280.
[49] *Pravda,* 22 October 1961.
[50] *Khrushchev Remembers,* p. 250. On the tapes from which his recollections were taken, Khrushchev, in discussing Rodionov and his fate, stated that "Zhdanov had worked many years in the Gorkiy region and he knew these cadres and therefore he, obviously, promoted them" (Russian-language transcripts, pt. I, p. 984).
[51] *Khrushchev Remembers,* p. 257. In the taped version of Khrushchev's recollections he explained: "Kuznetsov, you know, he was related to Kosygin. The wives of Kuznetsov and Voznesenskiy, that is, and of Kosygin—they had some sort of family

mittee in October 1938. By early 1939 he had been promoted to a Moscow job, people's commissar for the textile industry. In 1940 Kosygin became USSR deputy premier and in 1943 he became RSFSR premier, while remaining USSR deputy premier. In 1946 he became a candidate member of the Politburo, and sometime during 1948 he was promoted to full membership.[52] According to *Khrushchev Remembers*, Kosygin lost all his posts after the Leningrad Case, but by some stroke of luck he survived.[53] Actually, despite Khrushchev's statement, Kosygin officially remained a Politburo member and deputy premier until 1952–53. He also retained a ministerial post. On 28 December 1948 Kosygin was transferred from finance minister to head the Ministry for Light Industry,[54] which had just absorbed the Textile Industry Ministry.

The Fall of Voznesenskiy

The most important victim of the Leningrad Case was Politburo member, Deputy Premier and Gosplan Chairman N. A. Voznesenskiy, but his fall was precipitated partly by issues different from those that figured in the demise of the plot's other victims. Especially crucial to the fall of Voznesenskiy was the debate in economics, which started with the 1947 criticism of Professor Ye. S. Varga's book and continued through the publication of Stalin's economic theses in 1952. Voznesenskiy's downfall may have been inevitable

---

ties. Not that they were cousins really—something like that, I don't know, they were somehow related. And they were closely related, you know, by their wives" (Russian-language transcripts, pt. I, p. 826).

[52] Whether Kosygin was promoted from candidate member to full member of the Politburo before or after Zhdanov's fall remains impossible to determine. His biographies only list 1948 and scrutiny of press identifications and leadership lineups during 1948 fails to give a definitive answer. Kosygin clearly was still a candidate member in early 1948, since at the May Day parade and in other early 1948 lineups and listings of leaders, Kosygin always followed Politburo candidate member Shvernik. The first sign that Kosygin might be a full member occurred at the Air Day parade (25 July), when Kosygin stood among full members (seventh, ahead of Politburo member Voznesenskiy, as well as candidate member Shvernik). However, in describing the leaders ascending to the tribune to review the parade, *Pravda* listed Kosygin tenth—after Shvernik, as before. Moreover, listings and photos for the rest of 1948 (and even in 1949) almost invariably had Kosygin after Shvernik, suggesting that he was still only a candidate member.

[53] *Khrushchev Remembers*, p. 257. The taped version of Khrushchev's recollections does not say that he lost all his posts but that he was removed and then was appointed to "some sort of post of minister" (Russian-language transcripts, pt. I, p. 826).

[54] *Vedomosti verkhovnogo soveta SSSR*, 9 January 1949.

once his erstwhile patron Zhdanov had lost favor, but this was not apparent at the time. In fact, Voznesenskiy still appeared to be in good standing in late 1948, and his ouster in March 1949 apparently came as a surprise both to him and to his associates.

Voznesenskiy had enjoyed a spectacular rise during the early 1940s, clearly benefiting from Stalin's favor. He had become a candidate member of the Politburo on 21 February 1941, first deputy premier in charge of the Council of People's Commissars' Economic Council on 10 March 1941, and deputy chairman of the powerful State Defense Committee during World War II. Khrushchev stated that as first deputy premier Voznesenskiy often substituted for Stalin in presiding over the sessions of the Council of Ministers.[55] In October 1946, according to Khrushchev's secret speech, Voznesenskiy, still only a Politburo candidate member, became a member of the exclusive six-member "Politburo commission for foreign affairs"—the inner Politburo—and at the February 1947 CC plenum he was elected to full membership in the Politburo.

In late 1947 there were more signs of Stalin's trust in Voznesenskiy. On 6 August 1947, according to V. V. Kolotov (who was head of Voznesenskiy's secretariat in Gosplan and later his secretariat in the Council of People's Commissars and the Council of Ministers for over ten years), the CC and Council of Ministers approved Voznesenskiy's proposal to begin preparing a twenty-year plan, and soon Voznesenskiy's request that the Institute of Economics be subordinated directly to Gosplan was also approved.[56] In September 1947, after holding Voznesenskiy's manuscript on the

---

[55] *Khrushchev Remembers*, p. 251. Khrushchev presumably meant the Council of People's Commissars. When the Council of People's Commissars became the Council of Ministers in March 1946, the title "first deputy premier" was dropped and Voznesenskiy, like First Deputy Premier Molotov, thereafter was simply "deputy premier."

[56] Kolotov's biography of Voznesenskiy in the June 1974 *Znamya*, pp. 133–34. Kolotov has authored a number of versions of Voznesenskiy's career. Along with the Leningrad journalist G. A. Petrovichev, he turned out a forty-seven-page booklet on Voznesenskiy (*N. A. Voznesenskiy: biograficheskiy ocherk* [Moscow: Politizdat], signed to press 26 October 1963), based on Kolotov's knowledge and on information from Voznesenskiy's wife and sister, as well as various Voznesenskiy colleagues. Published under Khrushchev, this version had a distinct anti-Stalin slant, indicating that Voznesenskiy had acted independently of Stalin in pushing for a long-term plan, stating that Stalin had turned against Voznesenskiy because of the success of Voznesenskiy's book, and telling of Voznesenskiy's 1949 firing and of Beriya and Malenkov organizing the Leningrad Case against Voznesenskiy, Kuznetsov, and

"War Economy of the USSR in the Period of the Fatherland War" for almost a year, Stalin returned it to Voznesenskiy with his approval.[57] The book was published in a mass edition in December 1947 and was awarded a Stalin Prize in mid-1948.[58]

Voznesenskiy appears to have used his power to establish himself as the undisputed authority in the field of economics, purging all competing authorities. He took control over the economic institutes and used his book to discredit the leading economists, notably Varga, and to break up the establishment that dominated economic science (see Chapter 2).

The grounds for Voznesenskiy's downfall appear to have developed sometime in late 1948 or early 1949, as Stalin became distrustful and jealous. Voznesenskiy's widely publicized book had

---

Rodionov. Kolotov also wrote about Voznesenskiy in the June 1963 *Voprosy istorii KPSS*, contending that Voznesenskiy's book had incited Stalin's jealousy and had led to his 1949 purge.

The marking of Voznesenskiy's sixtieth birthday in late 1963 provided the occasion for publishing the most informative versions of Voznesenskiy's fall ever to appear. In a 1 December 1963 *Izvestiya* article, Petrovichev quoted Kolotov on details of Stalin's March 1949 firing of Voznesenskiy (see below), and the 1 December 1963 *Sovetskaya Rossiya* carried excerpts from the Kolotov-Petrovichev booklet. An article by Kolotov in the 30 November 1963 *Literaturnaya gazeta* included details not even contained in the booklet, notably the revelation that in 1949 Beriya had already put Voznesenskiy and other Gosplan leaders on trial for allegedly losing secret documents, but that this first case had failed (see below).

Khrushchev's fall and the end of revelations about crimes under Stalin drastically altered the portrayal of events in later biographies of Voznesenskiy. The eighty-page biography of Voznesenskiy by Kolotov published in the June 1974 *Znamya* included some details of Voznesenskiy's relations with Stalin but nothing at all about his 1949 removal. A longer (350-page) biography of Voznesenskiy by Kolotov was issued in May 1974 and, with slight changes, reissued in 1976. These books devote only two vague paragraphs to Voznesenskiy's fate, noting that he was "slandered by political careerists in connection with the so-called 'Leningrad Case'" and removed from all posts in March 1949.

Kolotov clearly has a fuller version of the Voznesenskiy story, including details as yet unpublished. Roy Medvedev (*Let History Judge* [New York, 1971], p. 481) cited a story from an unpublished manuscript by Kolotov entitled *Ustremlennyy v budushcheye* [Directed to the future]—the same title used for the 1974 *Znamya* account. The story concerned Voznesenskiy's refusal to sign a list of persons condemned to death which had been prepared by Beriya.

[57] Kolotov relates how Voznesenskiy showed him the manuscript with Stalin's corrections and initials on it (*Znamya*, June 1974, p. 135). CC Secretary L. F. Ilichev at the 1961 CPSU congress also noted that Stalin had personally edited Voznesenskiy's book (*Pravda*, 26 October 1961).

[58] The book was published in 250,000 copies, according to the *Vestnik akademii nauk SSSR*, no. 3, 1948, p. 57. The Stalin Prize was announced in the 30 May 1948 *Pravda*.

made him the economic theorist of the regime, a role Stalin himself apparently cherished. G. M. Sorokin, a subordinate of Voznesenskiy at Gosplan and now director of a prominent economic institute, wrote in *Pravda* (1 December 1963) that Voznesenskiy's book had "found a wide circle of readers, but its popularity had evoked the dissatisfaction of Stalin, who considered himself the lawgiver in the field of theory." Similarly, an article on Voznesenskiy's anniversary in the December 1963 *Planovoye khozyaystvo* declared that his book had upset Stalin because the latter "considered himself the sole authority in theory." Kolotov's 1963 booklet about Voznesenskiy also stated that the book had aroused Stalin's jealousy and cited as evidence statements made at the 1961 CPSU congress by L. F. Ilichev, the head of Agitprop.[59] Ilichev, who was deputy head of Agitprop at the time of Voznesenskiy's purge, declared that even though Stalin had edited the book, this "theoretician-god" grew dissatisfied when Voznesenskiy's book "became popular among economists and students" and when "some economists even zealously lauded it and found something new and creative in it."[60]

Stalin's dissatisfaction may have been intensified by the fact that in November 1948, buoyed by his first successful book, the hardworking and ambitious Voznesenskiy began work on a second book, a basic theoretical work entitled the *Politicheskaya ekonomiya kommunizma* (The political economy of communism). Even after being dismissed from all his posts in March 1949, Voznesenskiy continued work on this new book,[61] and when he was arrested in October, the manuscript was seized and destroyed.[62] In July 1949 a CC decree censured Agitprop's leaders for having "fawningly" praised Voznesenskiy's first book and then recommending it as a textbook.[63] In 1952 Stalin unveiled his own theoretical work on the economics of socialism, *Economic Problems of Socialism in the USSR*, which stimulated massive acclaim for Stalin as a great theoretician and economist. Khrushchev stresses that in 1951–52 Stalin was very concerned that he be regarded as the one and only economic theoretician.[64]

[59] Kolotov, N. A. *Voznesenskiy: biograficheskiy ocherk*, pp. 38–39.

[60] *Pravda*, 26 October 1961.

[61] According to 1 December 1963 Voznesenskiy anniversary articles by G. Petrovichev in *Izvestiya* and G. Sorokin in *Pravda*.

[62] Petrovichev in the 1 December 1963 *Izvestiya* and Kolotov in the 30 November 1963 *Literaturnaya gazeta*, and the 1963 Kolotov-Petrovichev booklet, p. 46.

[63] According to Suslov in the 24 December 1952 *Pravda*.

[64] *Khrushchev Remembers*, p. 272.

Stalin's turning against Voznesenskiy has been blamed mainly on Beriya, who apparently had motives for intriguing against Voznesenskiy. G. Petrovichev's anniversary article on Voznesenskiy (*Izvestiya,* 1 December 1963) blamed Voznesenskiy's demise on "Beriya's malicious slanders." In the 1 December 1963 *Pravda,* Voznesenskiy's former subordinate Sorokin blamed his fall on the Leningrad Case, which had been fabricated by Beriya and Malenkov. Kolotov's 1963 booklet declared that Beriya, "with Malenkov's help," organized the Leningrad Case against Voznesenskiy, Kuznetsov, and Rodionov.[65] Khrushchev blamed Beriya for leading the intrigue against Voznesenskiy and attributed Beriya's hostility to his fear of Voznesenskiy as a rival and to his frequent run-ins with Voznesenskiy over resource allocations.[66] The unpublished version of Kolotov's biography of Voznesenskiy apparently claimed that Voznesenskiy had provoked Beriya's retaliation by refusing to sign a list of persons condemned to death which had been prepared by Beriya.[67]

Although Beriya and Malenkov apparently had been intriguing against Voznesenskiy for some time, there were no signs in late 1948 that he was losing status. According to Kolotov, Voznesenskiy's ouster was sudden and unexpected. He writes that "on 5 March 1949 the unexpected, the unbelievable happened": Voznesenskiy was removed as deputy premier, head of Gosplan, and from all other state posts, and "several days later Stalin signed a decision expelling him from the Politburo of the CC VKP(b) and then expelling him from the CC."[68]

[65] Kolotov-Petrovichev, *N. A. Voznesenskiy,* pp. 46–47.
[66] Khrushchev's secret speech, p. S46, and *Khrushchev Remembers,* p. 252. According to *Khrushchev Remembers* (p. 251), Voznesenskiy had sought to redistribute resources more evenly, which meant taking money away from ministries under Beriya's supervision. The original tapes of Khrushchev's recollections have Khrushchev explaining that Voznesenskiy "often fought with Beriya, that is, when the plan was being formulated." Beriya demanded the "lion's share" of resources for ministries under his supervision and since there were not enough resources to satisfy everyone, Voznesenskiy wound up resisting Beriya's claims (Russian-language transcripts, pt. I, p. 991).
[67] Roy Medvedev (*Let History Judge,* p. 481) related this story of Kolotov's, citing Kolotov's unpublished manuscript.
[68] Petrovichev in the 1 December 1963 *Izvestiya,* quoting Kolotov's 1963 reminiscences. Actually Petrovichev's quote is a paraphrase and elaboration of the statement in the 1963 Kolotov-Petrovichev booklet itself, which does not give the exact date or list his posts (p. 46). The date of the decree was also confirmed in the 15 March 1949 *Pravda* account of the USSR Supreme Soviet session, which confirmed a 5 March ukase releasing him as Gosplan chairman and a 7 March ukase releasing him as deputy premier.

Beriya quickly organized a criminal case against Voznesenskiy and put him on trial, but the first trial misfired. Kolotov, writing in the 30 November 1963 *Literaturnaya gazeta,* described how Beriya sent his agents into Gosplan to check the handling of secret documents and after a big investigation managed to "establish the loss" of some secret papers. Kolotov relates that "on Beriya's order, a provocatory court case was immediately contrived, as a result of which USSR Gosplan Chairman N. A. Voznesenskiy was brought to trial." In order to mask the fact that this was simply a plot to get rid of Voznesenskiy, Voznesenskiy's assistants—first deputy Gosplan chairman A. D. Panov, deputy chairman A. V. Kuptsov, department chief S. K. Belous, and cadres department chief P. A. Oreshkin—were put on trial with him. However, the case was weak and on the first day of the trial Voznesenskiy, according to Kolotov, so effectively disproved the charges against him that Beriya dropped them and simply had the court convict the other four defendants, who received terms of four to six years in prison.

Having fought off the first case against him, Voznesenskiy stayed home and, until his arrest in late 1949, continued working on his new book, which, according to Kolotov, he finished in October 1949. Voznesenskiy was in limbo throughout this period. In the 30 November 1963 *Literaturnaya gazeta,* Kolotov, citing Voznesenskiy's wife, says that Voznesenskiy repeatedly sent letters to Stalin asking for work, and Khrushchev recalls that even after the ouster, Stalin allowed Voznesenskiy to come to his dinners and occasionally considered appointing him to some lesser post.[69] But Voznesenskiy was eventually arrested (around October 1949), tried in the Leningrad Case, and sentenced to death.[70]

As if reflecting Malenkov's involvement in Voznesenskiy's fall, some of Malenkov's men were installed in Gosplan's leadership after Voznesenskiy's ouster. M. Z. Saburov, who later appeared to have been close to Malenkov and who was ousted with Malenkov in the 1957 purge of the "Anti-Party Group," became Gosplan chairman on 5 March 1949.[71] Ye. Ye. Andreyev, long one of Malenkov's close

[69] *Khrushchev Remembers,* p. 251.

[70] Ibid., p. 256.

[71] *Leningradskaya pravda,* 15 March 1949. Saburov had replaced Voznesenskiy as Gosplan chairman in March 1941, when Voznesenskiy rose to first deputy premier, but surrendered this post in 1944 or 1945.

assistants in the CC cadres apparat, became secretary of Gosplan's party committee.[72]

Thus, 1949 saw the eradication of the Zhdanov faction in the party and government, along with the vicious campaigns against liberals and moderates in ideology and science. But no sooner had Stalin allowed, or himself organized, these purges than he began intriguing against his victorious deputies, Malenkov and Beriya, as well.

[72] Andreyev was last identified as deputy chief of the CC Cadres Administration in the 4 December 1947 *Vechernyaya Moskva*. The 5 February 1953 *Vechernyaya Moskva* identified him as secretary of Gosplan's party committee.

CHAPTER 5

Stalin's Last Years and the Succession —1950–55

The 1948–49 triumph of Malenkov, Beriya, and the dogmatists, although causing severe damage, did not end political maneuverings and establish these figures in permanent and indisputable control. Stalin was himself responsible for the defeat of moderation and, true to his style, he soon began to place limits on Malenkov and Beriya and to intrigue against the victorious dogmatists. In the last months before his death, Stalin began preparations for a new purge, both of party leaders and also of experts in the social sciences, especially economic science. The new purge probably would have swept away many of the purgers of 1949, but Stalin's death prevented the implementation of this new campaign. Recognizing the unpopularity and destructiveness of Stalin's most recent excesses—some of which had been aimed at them—his immediate successors Malenkov and Beriya quickly adopted notably less hard-line positions in foreign and domestic policy, repudiating some of Stalin's more extreme acts. But the surviving Zhdanovites sided with Khrushchev, the new party first secretary, against Malenkov and Beriya, and Khrushchev eventually bested Malenkov and became the hero of the moderates by openly denouncing Stalin and his excesses.

Leadership Maneuverings

With the purge of Zhdanov and his clique, Malenkov and Beriya clearly became Stalin's most powerful deputies. Stalin's concern about their power was manifested in his decision to bring Khrushchev into the Moscow leadership to serve as a check on

Malenkov, as well as by his 1951–53 moves to undermine Beriya. Stalin also turned increasingly to his personal secretary, A. N. Poskrebyshev, and to younger, junior figures whom he could use against his colleagues in the Politburo. His planned purge of Beriya and most of the other senior Politburo members failed to reach fruition, however, because of Stalin's death in March 1953.

Stalin's moves to check the growing power of Malenkov began as early as December 1949, when he had Ukrainian First Secretary and Politburo member N. S. Khrushchev transferred to Moscow to serve as CC secretary and Moscow first secretary. Khrushchev quickly emerged as a competitor for Malenkov, especially on the agricultural front. In his 1970 recollections, Khrushchev declared that he soon realized that his presence in Moscow "got in the way of Beriya and Malenkov's plans," and "I even began to suspect that one of the reasons Stalin had called me back to Moscow was to influence the balance of power in the collective and to put a check on Beriya and Malenkov. It seemed sometimes that Stalin was afraid of Beriya and would have been glad to get rid of him but didn't know how to do it." He added that "I was constantly running up against Beriya and Malenkov."[1]

By early 1950 Malenkov and Khrushchev had pushed the Politburo's other agricultural authority, Andreyev, out of any meaningful role in agricultural leadership. Andreyev, the chairman of the Kolkhoz Affairs Council, was publicly humiliated and discredited when an unsigned *Pravda* article (19 February 1950) attacked "the incorrect positions ... of Comrade A. A. Andreyev" on using small teams (links) rather than large brigades in agricultural production. A 25 February letter by Andreyev, printed in the 28 February 1950 *Pravda,* admitted that the criticism was correct and that his views on links were "mistaken." Khrushchev added his own criticism of links, but without naming Andreyev, in a March 1950 speech published in the 25 April 1950 *Pravda,* and Malenkov also attacked the small team concept in his report to the October 1952 party congress.

Andreyev remained chairman of the Kolkhoz Affairs Council until its eventual abolition in 1953,[2] but Malenkov soon strengthened his control over other agricultural posts. A. I. Kozlov,[3] a close associate

---

[1] *Khrushchev Remembers,* p. 250.
[2] See Chapter 1.
[3] See Chapter 1.

of Malenkov, was already head of the CC Agriculture Section, and on 27 October 1950 Ponomarenko, a Malenkov protégé and CC secretary, was named minister of procurements,[4] replacing B. A. Dvinskiy, who had been appointed in the mid-1940s while Andreyev had charge of agriculture.[5]

Although Malenkov was thus in a strong political position in the formulation of agricultural policy, Khrushchev had little respect for Malenkov's ability to lead this sector[6] and he soon moved in with his own proposals. In speeches on 16 March and 31 March 1950 (published in the 25 April 1950 *Pravda*), Khrushchev proposed consolidating small kolkhozes and building new, modern villages. Later, in a 18 January 1951 speech printed in the 4 March 1951 *Pravda*, he specifically proposed the construction of rural settlements with modern conveniences—to be called "agro-towns" or, as Khrushchev preferred, "kolkhoz settlements"—to replace the present villages.

Malenkov had become involved in agricultural leadership in 1946 and had received public credit for the 1947 harvest.[7] Clearly regarding himself as the Politburo's agricultural spokesman, Malenkov did not appreciate Khrushchev's growing influence in this field and set out to discredit Khrushchev's initiatives. The day after Khrushchev's "agro-towns" proposal appeared in *Pravda*, the same newspaper printed an editorial note stating that the article had been published only for "discussion" purposes, in other words, it did not represent government policy. At the 1961 party congress, CC Secretary L. F. Ilichev said that Malenkov had played a key role in this attack on Khrushchev. Ilichev described how Khrushchev's article had angered Stalin, who in turn accused Ilichev, then chief editor of *Pravda*, of "political immaturity" for printing it and ordered him to print a "correction." Then, on the instigation of Malenkov, a secret CC letter was prepared condemning Khrushchev's article as "anti-Marxist" and "mistaken."[8]

---

[4] Ponomarenko retained his post as CC secretary.

[5] Dvinskiy was appointed in 1944, according to his obituary in the 11 June 1973 *Pravda*.

[6] In his recollections, Khrushchev characterized Malenkov as "totally incompetent" in agriculture (*Khrushchev Remembers*, p. 236). In his 26 April 1958 Kiev speech and his December 1958 plenum speech, Khrushchev assailed Malenkov, whom he cited as responsible for agriculture in the early 1950s, for covering up the shortage of grain in order to conceal his own failures in managing agriculture (Khrushchev's speeches, vol. 3, pp. 203, 345).

[7] See the speech nominating Malenkov for the Moscow soviet in the 23 November 1947 *Pravda* (for more, see Chapter 1).

[8] *Pravda*, 26 October 1961. In his speech the following day at the 1961 congress, P.

Beriya appeared to join Malenkov in undermining Khrushchev on this issue. In his recollections, Khrushchev states that Beriya was then in charge of construction for the CC and opposed his ideas on prefabricated concrete construction.[9] Beriya's men in the Transcaucasus publicly joined in the campaign to discredit Khrushchev also. In his report to the March 1951 congress of the Armenian party, Armenian First Secretary G. A. Arutinov[10] ridiculed "some comrades" for urging resettlement from small villages into "agro-towns," or "kolkhoz settlements," thereby distorting the idea of consolidating small kolkhozes.[11] In his address to the May 1951 Azerbaydzhani party congress, Azerbaydzhan First Secretary M. D. Bagirov, one of Beriya's closest party protégés, assailed the "incorrect proposition" that villages should be merged into one settlement per kolkhoz and attacked attempts to force consolidation or turn it into a "campaign."[12]

Moreover, in his keynote report to the October 1952 Nineteenth Party Congress, Malenkov ridiculed Khrushchev's agricultural proposals, declaring that "some of our leading officials" erred in suggesting resettlement from villages to big "kolkhoz settlements" and "agro-towns." Malenkov explained that the mistake of "these comrades" was to forget about production—the main task in agriculture—and focus instead on improving the farmers' standard of living.[13]

Thus, in the 1951–52 maneuverings Malenkov still managed to remain ahead of Khrushchev and other top Stalin deputies. He was the only leader allowed to be Stalin's deputy in both party and government (CC secretary and deputy premier) and was ranked third

---

A. Satyukov followed up on Ilichev's story, declaring that Shepilov, on the orders of Malenkov and Molotov, had participated in writing the 1951 letter condemning Khrushchev's article (*Pravda*, 27 October 1961). The 1980 CPSU history mentions an April 1951 CC letter "On the Tasks of Kolkhoz Construction in Connection with Consolidation of Small Kolkhozes," which insisted that "the decisive role" in agriculture belongs to production and that attempts to switch resources from production to improving living standards were "premature" (*Istoriya kommunisticheskoy partii sovetskogo soyuza*, vol. 5, bk. 2, p. 346). This same volume of the CPSU history (p. 637) mentions a 2 April 1951 CC decree "On the Tasks of Kolkhoz Construction in Connection with the Consolidation of Kolkhozes."

[9] *Khrushchev Remembers: The Last Testament* (Boston, 1974), p. 96.

[10] Or Arutyunov (newspapers in Armenia spelled his name "Arutyunov" while those outside Armenia spelled it "Arutinov").

[11] *Kommunist*, 21 March 1951.

[12] *Bakinskiy rabochiy*, 26 May 1951.

[13] *Pravda*, 6 October 1952.

after Stalin and Molotov. His status as the closest thing to an heir apparent was reinforced when he was chosen to deliver the main address at the Nineteenth Party Congress. Moreover, more than any other Stalin deputy it was Malenkov who was able to get those who were clearly his friends elected to the leadership at the end of the congress. Malenkov's ally Ponomarenko was among the ten secretaries elected and the new thirty-six-man Presidium included Ponomarenko, as well as Gosplan Chairman M. Z. Saburov and Leningrad First Secretary V. M. Andrianov, the latter two having replaced the purged Zhdanovites Voznesenskiy and Popkov.

At the same time, however, Stalin clearly limited Malenkov's power and a number of Stalin's acts appeared detrimental to his position. He did not allow Malenkov any hand in the purge he was preparing, but instead appointed new people loyal to himself, not to Malenkov, to conduct it (see below). In fact, Ignatyev, the new MGB chief, had been associated with Zhdanov's faction and probably had poor relations with Malenkov. Moreover, Malenkov appears to have lost control over cadres. Whereas Malenkov protégé Shatalin had apparently headed the cadres department in the early 1950s,[14] N. M. Pegov had become cadres chief[15] by the time of the congress, and at its conclusion Chelyabinsk First Secretary A. B. Aristov was brought in to take over this post[16] and Shatalin was demoted to deputy head.[17] Malenkov acted to reverse this situation as soon as he succeeded Stalin as party leader in March 1953. In the 6 March 1953 reorganization, carried out while Malenkov was first secretary, Shatalin became CC secretary for cadres. On 14 March Aristov was

[14] He was listed as head of an unnamed CC section in the 30 November 1950 *Vechernyaya Moskva*. As Shatalin was a longtime cadres specialist, this section presumably was the Party, Trade Union, and Komsomol Organs Section.

[15] Pegov's biography in the 1979 *Deputaty verkhovnogo soveta SSSR* listed him as head of the CC Light Industry Section and then of the Party, Trade Union, and Komsomol Organs Section, 1948–52, and then as CC secretary and candidate member of the CC Presidium from 1952. He was last identified as Light Industry Section head in the 27 January 1951 *Pravda*. He apparently gave up leadership of the cadres section when elected secretary at the end of the congress.

[16] Aristov was identified as CC secretary and head of the Section for Party, Trade Union, and Komsomol Organs 1952–53 in his biography in the second edition of the *Bolshaya sovetskaya entsiklopediya*, vol. 51. He was in Chelyabinsk until the congress.

[17] Shatalin was identified as deputy head of an unnamed section in the 5 February 1953 *Vechernyaya Moskva*. He was elected only a candidate member of the CC at the 19th Congress, but when Malenkov became top leader in March 1953, Shatalin was quickly raised to full CC member at a 14 March plenum (*Pravda*, 21 March 1953).

dropped from the Secretariat and exiled to the Far East.[18] Aristov became an ally of Khrushchev against Malenkov and later was made CC secretary for cadres under Khrushchev.

Also detrimental to Malenkov was the fact that Stalin allowed him no hand in selecting members of the new Presidium and in fact included some of Malenkov's obvious enemies:[19] Deputy Premier A. N. Kosygin, MGB chief S. D. Ignatyev, Belorussian First Secretary N. S. Patolichev, and Krasnodar First Secretary N. G. Ignatov. Kosygin, a onetime Zhdanov protégé, had barely avoided perishing in the 1949 purges. Ignatyev and Patolichev had taken part in Zhdanov's 1946–47 intrigues in the CC apparat against Malenkov, and Patolichev's antagonism toward Malenkov is clear from his autobiography.[20] Ignatov was a victim of one of Malenkov's earlier purges. At the 18th Party Conference in February 1941, Malenkov had delivered a harsh attack on a number of party and government officials, and Ignatov was one of those expelled from CC candidate membership for failing to measure up to his duties. He was then fired as Kuybyshev first secretary. Patolichev describes this purge in his autobiography, complaining that it had been unnecessarily harsh and singling out Ignatov for sympathy: "I especially felt for ... Ignatov." He had been criticized for "shortcomings in the leadership of agriculture" at a plenum before the conference, according to Patolichev, "but the criticism was unsubstantiated." "I recall how one of those doing the criticizing, someone holding a very high post in the party, asked Ignatov to stand up and show himself," a demand which, said Patolichev, "made a not very good impression" on the other *oblast* secretaries who were present. Patolichev notes that Ignatov was then fired as *oblast* first secretary and demoted to head of a section in the Orel *oblast* party committee.[21]

Though Malenkov remained high in Stalin's favor right down to Stalin's death, Beriya was not so lucky. Stalin's moves against Beriya

[18] Aristov was listed as chairman of the Khabarovsk executive committee 1953–54 in his biography in volume 51 of the *Bolshaya sovetskaya entsiklopediya,* 2d ed.

[19] In his recollections, Khrushchev tells of asking Malenkov (who was the Politburo's cadres specialist) whether he had had a hand in selecting the new Presidium and relates that Malenkov denied it, saying that "Stalin didn't even ask for my help nor did I make any suggestions at all about the composition of the Presidium." Khrushchev then queried other leaders and eventually could only guess that Kaganovich must have been the one who helped Stalin pick the names (*Khrushchev Remembers,* pp. 279–80).

[20] See Chapter 1.

[21] N. S. Patolichev, *Izpytaniye,* pp. 113–14.

began in late 1951 and grew more ominous in late 1952 and early 1953. In late 1951 Beriya's protégé Abakumov was dismissed as MGB chief and arrested. Stalin appointed CC official S. D. Ignatyev as the new minister for state security and Odessa First Secretary A. A. Yepishev as deputy minister.[22] The new MGB leaders were clearly hostile to Beriya. Ignatyev had been associated with Zhdanov's maneuvers against Malenkov in 1946–47. He had been made deputy chief of the new CC Checking Administration in 1946 under Patolichev and in early 1947 was sent to Belorussia to become Belorussian agriculture secretary during the purge of Malenkov's protégés Ponomarenko et al. In March 1948 Ignatyev was elected Belorussian second secretary (*Pravda,* 19 March 1948), but after Zhdanov's downfall he was removed and wound up as "CC plenipotentiary" for Uzbekistan.[23] Yepishev was a longtime protégé of Khrushchev. Khrushchev had named him secretary for cadres of the Ukrainian CC in mid-1946, but Yepishev became inactive after new Ukrainian First Secretary Kaganovich made L. G. Melnikov CC secretary in charge of organizational work in July 1947. After Khrushchev returned as Ukrainian first secretary in December 1947, Yepishev resumed an active role.[24] From January 1950 to September 1951 he was first secretary of the important Odessa party organization. As deputy minister of state security from 1951 to 1953, Yepishev was clearly an anti-Beriya element in the police apparat, and, as soon as Beriya regained control of the MGB upon Stalin's death, Yepishev was ousted. Yepishev was sent to Odessa and on 29 March 1953, *Pravda Ukrainy* reported his election as first secretary of Odessa. Later, after Khrushchev became CPSU party leader, he trusted Yepishev enough to name him political chief of the nation's

[22] The exact date of Ignatyev's appointment is unclear. According to Boris Nikolayevskiy, writing in the émigré newspaper *Sotsialisticheskiy vestnik* (January 1955), Abakumov was removed in November–December 1951. Nikolayevskiy states that Abakumov was seen at the 7 November 1951 parade and therefore his removal could not have occurred earlier. (*Pravda*'s accounts of the November anniversary ceremonies do not mention lower-ranking officials such as Abakumov, so his presence cannot be confirmed.) Yepishev, according to his biography in the third edition of the *Bolshaya sovetskaya entsiklopediya,* became deputy minister of state security in 1951. His release as Odessa first secretary was reported in the 4 September 1951 *Pravda Ukrainy.*

[23] So identified in *Pravda,* 24 February 1950.

[24] He attacked and removed the *oblast* first secretary at a Chernigov plenum (*Pravda Ukrainy,* 6 March 1948), addressed a republic conference of *oblast* cadres secretaries (*Pravda Ukrainy,* 24 April 1948), and spoke at a republic Komsomol plenum (*Pravda Ukrainy,* 9 July 1948).

armed forces (chief of the Main Political Administration) in 1962. In November 1951 Stalin opened an attack on Beriya's main power base in the party, the Georgian party organization. According to Khrushchev's secret speech, Stalin "personally dictated" November 1951 and March 1952 CC decrees that accused various Georgians of nationalism and treason.[25] In April 1952, Beriya's top protégés in the leadership—including First Secretary K. N. Charkviani—were replaced with others not close to Beriya. A. I. Mgeladze, the new first secretary, began a massive purge of the party[26] and, along with the Georgian minister of state security, fabricated documents "proving" that various of Beriya's local protégés were guilty of nationalism.[27] Khrushchev described this affair, which he called the "Mingrelian Case," in his secret speech. The falsified documents supposedly proved that there was a conspiracy among Mingrelians in Georgia to have the region secede from the Soviet Union.[28] As a result, Khrushchev continued, various Georgian leaders were arrested and "thousands of innocent people fell victim to willfulness and lawlessness." Beriya was not named directly, but he himself was a Mingrelian and it was his supporters who were being accused.[29] The net effect was to deprive Beriya of one of his main bases of support. After Stalin's death, Beriya quickly purged the new Georgian leaders and exposed the accusations as fabrications.[30]

The threat to Beriya became very obvious in late 1952 and early 1953, when Stalin had a group of Jewish doctors arrested and accused of murdering Zhdanov and plotting to murder other Kremlin leaders. The investigation of the Doctors' Plot was conducted under Stalin's personal direction by Ignatyev, the new MGB chief.[31] On 13 January 1953, *Pravda* announced that the doctors had been

[25] Khrushchev's secret speech, p. S47.
[26] For detailed descriptions of the purge, see John Ducoli, "The Georgian Purges (1951–53)," *Caucasian Review*, no. 6, 1958, pp. 54–61, and Robert Conquest, *Power and Policy in the U.S.S.R.* [New York, 1961], pp. 129–53.
[27] This fact was revealed during the April 1953 purge of Mgeladze and his confederates—see the speech of the new Georgian premier, V. M. Bakradze, in the 16 April 1953 *Zarya vostoka*. Bakradze said that Mgeladze had participated actively in the arrests. These charges were repeated in a 21 April 1953 *Zarya vostoka* editorial.
[28] Khrushchev's secret speech, pp. S46–47.
[29] In his taped recollections, Khrushchev declared that Stalin "raised the so-called Mingrelian Case" and that "this was an action directed by Stalin against Beriya, because Beriya was a Mingrelian" (Russian-language transcripts, pt. I, p. 154). Also see *Khrushchev Remembers*, p. 312.
[30] *Zarya vostoka*, 14, 16, 17, 18, 22, and 29 April 1953.
[31] Khrushchev in his secret speech (p. S49) and in his recollections (*Khrushchev*

arrested and were under investigation, and in a separate, front page editorial criticized the state security organs for not uncovering the conspiracy "in good time"—a pointed threat to Beriya.

Stalin also began moving against several other Politburo members during this period. Andreyev was already in public disfavor by 1950 and, according to Khrushchev, Stalin soon turned against Voroshilov as well, suspecting him of being an "English spy."[32] Then Stalin began attacking Molotov and Mikoyan, and in October 1952, at the first plenum after the 19th Congress, he openly accused them of being some sort of enemy agents.[33] In his secret speech, Khrushchev stated that if Stalin had lived several months longer, Molotov and Mikoyan probably would not have survived.[34] He mentioned that Stalin's suspicions about Molotov being an American agent appeared to stem from Molotov's visit to the United States,[35] and there have been some hints that Stalin was preparing a case against Molotov.[36]

Stalin gradually excluded these men from leadership meetings, even though there were virtually no changes in formal membership in the Politburo, the Secretariat, or the leadership of the Council of Ministers from 1949 to 1952. Khrushchev stated that Stalin banned Andreyev and later Molotov and Mikoyan from Politburo meet-

---

Remembers, pp. 286–87) describes Stalin ordering Ignatyev to beat confessions out of the doctors.

[32] Khrushchev's secret speech, pp. S62–63; and *Khrushchev Remembers,* pp. 281, 308.

[33] *Khrushchev Remembers,* pp. 278, 281, 309–10. In his taped recollections, Khrushchev notes how astonished he was at the post-congress plenum when Stalin attacked Molotov and Mikoyan. He indicates that Stalin had said such things in small groups but not at a formal meeting until then. He also notes that Molotov and Mikoyan spoke at this plenum (Russian-language transcripts, pt. I, p. 893).

[34] Khrushchev's secret speech, p. S63.

[35] *Khrushchev Remembers,* p. 309.

[36] Aleksandr Nekrich suggests that the early 1953 arrest of former ambassador to England I. M. Mayskiy and several other diplomats who had served in the Soviet Embassy in London during the war may have been intended to produce the evidence against Molotov. Mayskiy was ambassador in 1942 when Foreign Affairs Commissar Molotov visited England. Nekrich learned that some of those arrested were questioned about Molotov's visit. Mayskiy himself, accused of being an English spy, was apparently not asked about the visit, judging from what he told his friend Nekrich, but this may have been because his arrest occurred only in late February and Stalin's death aborted pursuance of this aspect of the investigation (Nekrich, *Otreshis ot strakha,* pp. 131–33; and Nekrich's article "The Arrest and Trial of I. M. Maysky" in the summer 1976 *Survey,* nos. 100–01, pp. 313–20). Khrushchev's mention of Stalin's suspicions over Molotov's visit to the United States appears to lend some credence to this idea.

ings,[37] and that Voroshilov and Kaganovich rarely were allowed to attend either.[38] Moreover, according to Khrushchev, Stalin began to hold Politburo meetings only occasionally,[39] and after the 19th Congress the leadership was reduced to virtually five men: Stalin, Malenkov, Beriya, Khrushchev, and Bulganin.[40]

In his moves against various Politburo members, Stalin acted through his personal secretariat headed by Poskrebyshev and through second-level leaders whom he had recently appointed: S. D. Ignatyev, who became MGB chief in late 1951; N. M. Pegov, who became head of the CC Section for Party, Trade Union, and Komsomol Organs shortly before the congress and apparently conducted the organizational work for the congress;[41] A. B. Aristov, who became CC secretary and head of the Section for Party, Trade Union, and Komsomol Organs at the conclusion of the congress in October 1952;[42] and N. A. Mikhaylov, who became CC secretary and Agitprop head in October 1952.[43] These men appeared to have few ties to Malenkov, Beriya, or other Stalin deputies and apparently were intended to be Stalin's personal agents. That these men were being

[37] Khrushchev's secret speech, pp. S62–63; and *Khrushchev Remembers*, pp. 281, 309.

[38] *Khrushchev Remembers*, pp. 281, 308.

[39] Khrushchev's secret speech, p. S61.

[40] *Khrushchev Remembers*, pp. 280–82.

[41] Pegov's biography in the 1979 *Deputaty verkhovnogo soveta SSSR* listed him as head of the Light Industry Section, then head of the Section for Party, Trade Union, and Komsomol Organs 1948–52, then CC secretary and candidate member of the CC Presidium from 1952. (All previous biographies of Pegov, for example, in the *Bolshaya sovetskaya entsiklopediya* yearbooks, had only identified him as "in the CPSU CC apparat 1948–52.") Since Pegov had been identified as head of the Light Industry Section in the 27 January 1951 *Pravda*, he would have become cadres chief sometime in 1951 or early or mid-1952. In this capacity he would have been in charge of preparations for the 19th Congress. This likelihood is borne out by the fact that Pegov was elected chairman of the Credentials Commission at the congress (*Pravda*, 6 October 1952), a post usually held by the head of the cadres department. (This has led Robert Conquest, in his 1961 book *Power and Policy in the U.S.S.R.* [p. 157], to speculate that Pegov was the cadres chief at the time of the congress, although the only piece of direct evidence—the *Deputaty* identification—appeared first in 1979.)

[42] According to his biography in volume 51 of the *Bolshaya sovetskaya entsiklopediya* (2d edition, printed 1958), Aristov was CC secretary and head of this section, 1952–53. His election as secretary at the CC plenum at the close of the October 1952 congress was publicized, but his selection as section head was not.

[43] According to his biography in the *Bolshaya sovetskaya entsiklopediya* yearbook for 1966, Mikhaylov was CC secretary and Agitprop head 1952–53. As with Aristov, his appointment as section head was not made public at the time, although his election as secretary was. He was released as Komsomol first secretary in early November 1952 to take up his new post.

used by Stalin against his leading deputies is suggested by the retaliation against them after Stalin died. The two who had taken over cadres work from Malenkov's protégé Shatalin—Pegov and Aristov—apparently were punished on Malenkov's initiative: Pegov was dropped as CC secretary on 6 March 1953 and demoted to the powerless post of secretary in the Presidium of the Supreme Soviet,[44] while Aristov was ousted as CC secretary on 14 March and exiled to the Far East.[45] Ignatyev, a participant in Stalin's plot against Beriya, was ousted in early April when Beriya exposed the Doctors' Plot as a fabrication and *Pravda*'s 6 April editorial on the case accused Ignatyev of "political blindness."[46] Mikhaylov was initially used by Malenkov against Khrushchev: when Khrushchev had to surrender his post as Moscow first secretary on 6 March, Mikhaylov was put in charge of this important organization. After gaining control over the party apparat, Khrushchev in early 1954 ousted Mikhaylov and put his own protégé, I. V. Kapitonov, in charge of the Moscow party organization.[47]

The CC Presidium elected at the 19th Congress to replace the old Politburo and Orgburo also reflected Stalin's drive against the old Politburo members. One Politburo member, Andreyev, was left out of the new Presidium entirely, while another, Kosygin, was demoted to a candidate member in the new Presidium. Even the Politburo members carried over to the new Presidium were swamped among its twenty-five members and eleven candidate members (for a list of members before and after the 19th Congress, see Appendix 1b).

Khrushchev indicates that the old Politburo members themselves viewed the new Presidium in this threatening light. As he later described it, Stalin himself selected the members of the new Presidium without consulting other Politburo members, even Malenkov and Khrushchev. Khrushchev characterized many of the new members as obscure even to the Politburo members themselves.[48] In his

[44] *Pravda*, 7 March 1953.

[45] His biography in volume 51 of the second edition of the *Bolshaya sovetskaya entsiklopediya* listed him as chairman of the Khabarovsk *kray* executive committee 1953–54. Khrushchev brought him back from exile and made him CC secretary again in 1955, after Malenkov had been defeated and forced to resign as premier.

[46] Ignatyev had been transferred to CC secretary on 6 March 1953, after Beriya took over the MVD and MGB. Ignatyev was dropped as secretary in early April (*Pravda*, 7 April 1953). Khrushchev later aided Ignatyev, making him first secretary of Bashkiria in early 1954 (*Pravda*, 26 February 1954).

[47] *Pravda*, 1 February 1954. Mikhaylov was sent off as ambassador to Poland and remained in second-rate jobs throughout Khrushchev's reign.

[48] *Khrushchev Remembers*, p. 279.

secret speech, he said: "Stalin evidently had plans to finish off the old members of the Politburo. He often stated that Politburo members should be replaced by new ones. His proposal, after the 19th Congress, concerning the election of twenty-five persons to the Central Committee Presidium, was aimed at the removal of the old Politburo members and the bringing in of less experienced persons so that these would extol him in all sorts of ways."[49]

The real leadership organ (the existence of which was not even publicly announced) was a nine-man "bureau" of the Presidium. According to Khrushchev, it included only Stalin, Malenkov, Beriya, Khrushchev, Voroshilov, Kaganovich, Saburov, Pervukhin, and Bulganin, thus excluding Politburo members Molotov and Mikoyan.[50] Furthermore, the new Presidium was never convened[51] and usually only five members of the bureau made policy decisions. The five-man Secretariat (Stalin, Malenkov, Khrushchev, Suslov, Ponomarenko) was expanded to ten at the end of the October 1952 congress, with the addition of Aristov, Mikhaylov, Pegov, Ignatov, and Brezhnev.

Stalin's new antagonism to Molotov, Mikoyan, Voroshilov, and Andreyev was apparently not made clear to party leaders outside the Politburo itself, since the republic party congresses held in September 1952 generally continued to rank the members of the old Politburo in an order that did not reflect Stalin's changing mood. The order used in electing honorary presidiums at local congresses was consistently: Stalin, Molotov, Malenkov, Beriya, Voroshilov, Mikoyan, Bulganin, Kaganovich, Andreyev, Khrushchev, Kosygin, and Shvernik.[52] Only the Georgian and Armenian parties appeared to reflect the changing situation regarding Molotov. Armenia elected Stalin, Malenkov, Beriya, and Mikoyan (but not Molotov) as honorary delegates to the 19th Congress, and included Stalin, Malenkov, Molotov, Beriya, and Mikoyan in the new Armenian CC—with Molotov listed after Malenkov (*Kommunist*, 23 September 1952). In Georgia, Molotov had been listed ahead of Malenkov and Beriya at occasions in early

---

[49] Khrushchev's secret speech, p. S63.
[50] *Khrushchev Remembers*, p. 281.
[51] Ibid., p. 281.
[52] Congresses in Georgia, Moldavia, Kirgizia, Latvia, and Turkmenia. (Other republic congresses only elected the Politburo "headed by" Stalin without listing the members.) Moreover, the Kirgiz congress, Moscow *oblast*, and Leningrad party conferences followed this order in electing all Politburo members as delegates from their area to the 19th Congress.

and mid-1952,[53] but by September he was being omitted from honorary lineups.[54] However, the honorary presidium at the Georgian congress followed the same order as elsewhere—probably reflecting Moscow's instructions on how to rank the Politburo, instructions perhaps intended to mask the changes in status.

The lineup on the opening day of the 19th Congress itself was a little more revealing, since it ranked Beriya and Mikoyan low, but at the closing session Beriya was back in his customary fourth position. *Pravda*'s order in listing leaders on 6 October and 15 October 1952, compared to republic listings in September, is illustrated in Chart C.

Chart C. Rankings of leadership, Republic and 19th Party Congresses

Republic congresses in September 1952	Opening day of 19th Congress (*Pravda*, 6 October 1952)	Closing day of 19th Congress (*Pravda*, 15 October 1952)
1. Stalin	1. Stalin	1. Stalin
2. Molotov	2. Molotov	2. Molotov
3. Malenkov	3. Malenkov	3. Malenkov
4. Beriya	4. Voroshilov	4. Beriya
5. Voroshilov	5. Bulganin	5. Voroshilov
6. Mikoyan	6. Beriya	6. Bulganin
7. Bulganin	7. Kaganovich	7. Kaganovich
8. Kaganovich	8. Khrushchev	8. Khrushchev
9. Andreyev	9. Andreyev	9. Andreyev
10. Khrushchev	10. Mikoyan	10. Mikoyan
11. Kosygin	11. Kosygin	11. Kosygin
12. Shvernik		12. Shvernik

Stalin's attack on Molotov, which, according to Khrushchev,[55] occurred at the 16 October postcongress plenum, was apparently reflected in Molotov's low standing at the ceremonies celebrating the October Revolution on 7 and 8 November 1952.[56] Though *Pravda* did not list the leaders in describing the ceremonies, its photos showed that Molotov was no longer at Stalin's side. At the 6

[53] The Tbilisi city party conference in *Zarya vostoka*, 13 and 15 April 1952, the Tbilisi *oblast* party conference in *Zarya vostoka*, 27 April, and Georgian Komsomol congress in the 16 May *Zarya vostoka*.

[54] In the election of the Tbilisi city party committee (*Zarya vostoka*, 9 September 1952), Tbilisi *oblast* party committee (*Zarya vostoka*, 13 September 1952), and Georgian CC (*Zarya vostoka*, 19 September 1952), only Stalin, Malenkov, and Beriya were included.

[55] Khrushchev's secret speech, p. S63.

[56] In listing the newly elected Presidium on 17 October, *Pravda* had given no clue to hierarchical standing, since it simply used alphabetical order, except for Stalin who was listed first.

November indoor ceremony, Beriya and Malenkov sat between Stalin and Molotov, while at the 7 November parade, Molotov stood well down in the rankings: Malenkov stood next to Stalin and the speaker, Marshal Timoshenko, followed by Beriya, Khrushchev, Kaganovich, Molotov, Shvernik, Pervukhin, Saburov, Mikoyan, Ponomarenko, Suslov, Shkiryatov, Aristov, Pegov, and Brezhnev.

Stalin's Economic Theses

After the 1949 crackdown, moderates understandably laid low and/or joined the orthodox chorus. The anticosmopolitanism drive, the Leningrad purge, and even the purge of anti-Lysenkoists were succeeded by new campaigns. Political figures such as Voznesenskiy and Kuznetsov, the Georgian leaders, and Jews suffered the worst treatment, including arrest and execution, whereas disgraced ideological officials and social scientists tended to be forgotten after an initial scourging. In late 1952, however, turmoil erupted anew in ideology and social sciences as part of a new secret maneuver by Stalin. Stalin published what was presented as a major theoretical and scientific work on the economic laws of socialism—his *Economic Problems of Socialism in the U.S.S.R.*—which quickly came to dominate ideological discussion. There ensued a new campaign of denunciations of economists and others for being out of step with Stalin's theses. New ideological leaders were appointed by Stalin and a major purge appeared in the making. As with his projected purge of the Politburo, however, this campaign was cut short by Stalin's death.

While the 1949 campaigns died down quickly, Stalin launched other crackdowns on seemingly innocent works of literature, culture and linguistics. The sharpest were the 1951–52 drive against local nationalisms, the execution of prominent Jewish writers in 1952, and the apparent preparation of a national pogrom in early 1953. It is hard to attribute these campaigns to anyone but Stalin. The 1951 attack on Ukrainian nationalism, for example, clearly caught even Melnikov, the pro-Russian Ukrainian first secretary, off guard. Melnikov had just finished praising Ukrainian writer Volodymyr Sosyura in a 15 June 1951 *Pravda* article, when *Pravda* began a series of attacks (16 June, 28 June, 2 July, and 7 July) on a seemingly innocuous Ukrainian poem that Sosyura had written seven years earlier, as well as on a new Ukrainian opera by Ukrainian Writers' Union Chairman Oleksandr Korniychuk and his wife. Stalin's personal sponsorship of the 1952–53 anti-Semitic campaign was clear from the fact that it ended abruptly upon his death and that his successors quickly repudiated the Doctors' Plot as a fabrication. Despite the turmoil at the top from 1950 to 1953, some moderates or former Zhdanovites escaped lightly, perhaps because they had residual support from Stalin. Kedrov and Varga continued to hold prominent jobs as members of editorial boards of leading journals, while Yuriy Zhdanov still headed the CC's Science Section.[57] Aleksandrov, perhaps now a protégé of Malenkov, remained director of the Institute of Philosophy until after Stalin's death.

A new chapter in the turmoil among ideologists and social scientists began with the publication of Stalin's economic theses *Economic Problems of Socialism in the U.S.S.R.* in the 3–4 October 1952 *Pravda*. Published on the very eve of the 19th Congress, the theses were clearly designed to help Stalin thoroughly dominate a congress at which he might only be able to deliver a short speech. Publication of the theses led to further exaltation of Stalin as a "genius" in economic theory and a "*korifey*" (coryphaeus, or outstanding figure) of science. Moreover, the appearance of the new work seems to have been related to his preparations for a purge of various political leaders. Exactly how the campaign based on Stalin's economic theses was intended to develop is unclear, because it, like his planned purge of the party, was cut short by his death.

Stalin's theses consisted of four articles, written at various times, on various pretexts, and on various subjects, but now published together and treated as a coherent manuscript.

The first and longest part was a collection of comments about a November 1951 debate over a draft of a textbook on political economy.[58] In his remarks, Stalin disagreed with those who con-

---

[57] He was identified as head of this section in the 16 October 1951 *Sovetskaya Belorussiya*. He still appeared to be in Stalin's favor in early 1953, when he wrote a 16 January 1953 *Pravda* article attacking deviations in natural sciences. (He was not identified in any post on this occasion, however.) Stalin's successors were less generous with Yuriy. By mid-1954 he had been exiled from Moscow and demoted to head of the science and culture section of the Rostov *oblast* party committee (so identified in the 3 July 1954 *Literaturnaya gazeta*).

[58] The November 1951 debate does not appear to have been reported in the economic journals. Stalin's comments, which are dated 1 February 1952, refer to an apparent transcript of the debate, however. The 1980 CPSU history reveals that the "preparation of a political economics textbook" had been started on the CC's

tended that economic laws could be changed by the state, that the "law of value" should regulate the economy and hence that the more profitable light industries should be given priority over the less profitable heavy industrial sector, and that Lenin's theses on the inevitability of wars between capitalist countries is outdated. Stalin mentioned Ostrovityanov, Leontyev, Shepilov, and Gatovskiy as among those participating in the debate. The second part, dated 21 April 1952, is an answer to economist A. I. Notkin, disagreeing with him on a number of points.

The third part, dated 22 May 1952, is an attack on "the errors of Comrade Yaroshenko L. D." As is clear from Stalin's comments, Yaroshenko participated in the November 1951 debate, but then had the temerity not only to write a letter to the Politburo disagreeing with the ideas expressed in the debate but also to contest Stalin's February 1952 remarks on the debate and to ask that he himself be entrusted to write a textbook on political economy. Stalin characterizes Yaroshenko's views as simply "un-Marxist," maintains that they are similar to Bukharin's, and disputes them at length. Although Stalin presents elaborate arguments against Yaroshenko's views, Khrushchev contends that there was no basic disagreement between the views of Stalin and Yaroshenko and that the complicated disagreements voiced by Stalin were contrived purely to discredit Yaroshenko personally.[59]

The last part, dated 28 September 1952, was an answer to a letter from economists V. G. Venzher and A. V. Sanina (his wife) which had proposed that state farm equipment be sold to kolkhozes. Accusing them of "serious theoretical errors," Stalin disagreed with the proposal as well as with their view that economic laws were subject to change by state organs.

Publication of Stalin's theses immediately led to feverish activity

---

"initiative," that 260 scholars had participated in the late 1951 debate on the outline, and that the Politburo on 16 February 1952 had created a commission to finish work on the outline (*Istoriya kommunisticheskoy partii sovetskogo soyuza*, vol. 5, bk. 2, pp. 252–53).

[59] Khrushchev relates that Stalin had found some of his own ideas in a book by an economist with a Ukrainian name (which he could not recall but who obviously was Yaroshenko), which had been boldly sent to the Politburo and which was called to Stalin's attention by Voroshilov in the summer of 1952. Stalin, infuriated, launched his denunciation of Yaroshenko, who was eventually expelled from the party and arrested. According to Khrushchev, he was released after Stalin's death (*Khrushchev Remembers*, pp. 271–72, 274). Yaroshenko's arrest and later release are also mentioned in *Politicheskiy dnevnik* II (Amsterdam, 1975), p. 621.

among economists and condemnation of various economists and other social scientists for holding mistaken views that conflicted with the theses. The scholarly council of the Institute of Economics held an expanded meeting on 4–5 November 1952 to discuss Stalin's theses and the errors of several of the institute's workers. Acting director F. V. Samokhvalov opened with an attack on errors by "former" institute worker I. A. Anchishkin and also on Ya. A. Kronrod, V. G. Venzher, and L. D. Yaroshenko. Venzher and Kronrod spoke and conceded some errors, as did Varga.[60] On 7–10 January 1953, the economics and law division of the Academy of Sciences also held a session on Stalin's theses and heard Institute of Economics director Ostrovityanov, Samokhvalov, A. M. Rumyantsev, and others attack economists L. Leontyev, I. Anchishkin, I. Gladkov, and L. Yaroshenko for "subjectivist-idealist" errors.[61] A January 1953 *Voprosy ekonomiki* editorial noted Suslov's criticism of *Voprosy ekonomiki* in a 24 December *Pravda* attack on Fedoseyev (see below) and admitted that *Voprosy ekonomiki* had erred in praising Voznesenskiy's "anti-Marxist" book. It also attacked Ostrovityanov, A. Leontyev, G. Kozlov, L. Gatovskiy, I. Gladkov, Ya. Kronrod, and *Voprosy ekonomiki*'s deputy chief editor, P. A. Belov, for praising Voznesenskiy's book. A 12 January 1953 *Pravda* article by CC official A. I. Sobolev[62] attacked Gladkov and Kozlov for "subjectivist" views and A. Leontyev for praising Voznesenskiy's book. The January 1953 *Vestnik* of the Academy of Sciences published a call by the academy's Presidium for improvement of the work of the Institute of Economics in the light of Stalin's theses and assailed the "subjectivist-idealistic, scholastic and other non-Marxist views" of Anchishkin, Venzher, Notkin, and Kronrod. At its annual meeting (30 January–2 February), the Academy of Sciences heard Yudin attack economists and philosophers for errors in a special report on Stalin's theses.[63]

In addition to having praised Voznesenskiy in his heyday, the most frequent deviation cited in the crackdown of late 1952 and early 1953 was that of holding the "subjectivist" view that economic laws could be altered or created by the state—a view contrary to Stalin's thesis that economic laws existed objectively and must be conformed

---

[60] *Voprosy ekonomiki*, December 1952, pp. 102–15.

[61] *Voprosy ekonomiki*, April 1953, pp. 78–95.

[62] A. I. Sobolev had been listed as "leader of a group of lecturers" in Agitprop in the 19 April 1952 *Pravda*.

[63] *Pravda*, 3 February 1953, and *Vestnik akademii nauk SSSR*, March 1953.

to. Those coming under attack included both "liberals" and "conservatives," suggesting that even those doing the attacking knew little about Stalin's real aim. Thus, while Venzher has been one of the most reform-minded economists both under Stalin and since, others accused of deviationism, such as Gladkov[64] and G. A. Kozlov,[65] were quite conservative. Those doing the attacking included conservatives like Yudin and Ostrovityanov, as well as Rumyantsev, who later developed a reputation as one of the more liberal economists.

Clearly, the publication of Stalin's theses was part of a larger design for a purge of social sciences, for at about the same time as the publication of the theses and the opening of the congress, Stalin appointed new people to most key ideological posts. N. A. Mikhaylov was named CC secretary and Agitprop head about October 1952 (see above), D. I. Chesnokov became coeditor of *Bolshevik* in November 1952[66] and head of an unnamed CC section,[67] D. T. Shepilov became chief editor of *Pravda*,[68] replacing Ilichev, and P. N. Pospelov became deputy chief editor of *Pravda*.[69] Suslov, who remained CC secretary and the top ideological official, Mikhaylov, and Chesnokov were named full members of the new CC Presidium and veteran dogmatist Yudin became a candidate member of the Presidium. Another member of the new group was A. M. Rumyantsev, director of the Ukrainian Institute of Economics, who was brought to Moscow sometime during 1952 and soon became head of the new Science and Culture Section of the CC.[70] He was

[64] See Chapter 2.

[65] Kozlov was named deputy director of the new Institute of Economics in 1948, after Varga's purge, and helped lead attacks on Varga. Identified as deputy director, he opened a 29–30 March 1948 institute debate on "non-Marxist" views of Varga, Trakhtenberg, and others (*Voprosy ekonomiki*, no. 2, 1948, p. 107). However, he was fired as deputy director in late 1950 after allowing publication of a book on economic crises by L. Mendelson which later (29 September 1950) was attacked in *Pravda* (see the November 1950 *Voprosy ekonomiki*, p. 108).

[66] So identified starting in *Bolshevik*, no. 21, 1952.

[67] So identified in the 5 February 1953 *Vechernyaya Moskva*.

[68] So identified in the 17 December 1952 *Izvestiya*.

[69] So identified in the 6 February 1953 *Vechernyaya Moskva*. He had been director of the Institute of Marx-Engels-Lenin.

[70] Rumyantsev's biography in the third edition of the *Bolshaya sovetskaya entsiklopediya* indicates that he was "head of the Science Section" from 1952 to 1955 and his biography in the Siberian journal *Eko* (no. 2, 1974, p. 42) lists him as "head of the Science and Culture Section" 1952–55. Aleksandr Nekrich (*Otreshis ot strakha*, pp. 78, 112) mentions Rumyantsev as being head of the Science Section as of 21 April 1953. Rumyantsev was clearly in a prominent position in Moscow by late 1952 in view of his election to the CC and to the commission to rewrite the Party Program at

elected a full CC member at the congress and named a member of the select commission to revise the Party program.[71] Moreover, some new bodies were created. An Ideological Commission was set up under the new Presidium, and though its purpose has never been made clear,[72] likely members were: M. A. Suslov, CC secretary and Presidium member, N. A. Mikhaylov, CC secretary, Presidium member and Agitprop head, D. T. Shepilov, chief editor of *Pravda*, D. I. Chesnokov, co-chief editor of *Bolshevik* and head of a CC section, S. M. Abalin, co-chief editor of *Bolshevik*, P. F. Yudin, Presidium member and academician, A. M. Rumyantsev, head of CC Science and Culture Section, K. A. Gubin, chief editor of *Izvestiya*.

This group acted as spokesmen for the campaign based on Stalin's theses. Suslov delivered a big attack in the 24 December 1952 *Pravda*; Chesnokov gave a speech on the theses at a 23–26 January 1953 Institute of Philosophy meeting[73] and wrote an article on Stalin in a January 1953 *Kommunist*;[74] Mikhaylov delivered the 21 January 1953 Lenin-day speech warning of spies and murderers and citing intensification of class war and the need for vigilance;[75]

---

the congress. The Science and Culture Section apparently replaced the Science and Vuzes (higher educational institutions) Section headed by Yuriy Zhdanov. Though it is not certain that the reorganization occurred before Stalin's death, it is probable, in view of the 1974 *Eko* identification of Rumyantsev and because a "Science and Culture Section" was already mentioned in the 21 April 1953 *Pravda*, both of which indicate that it occurred soon after Stalin's death at the latest.

[71] The eleven members included Presidium members Stalin, Beriya, Kaganovich, Kuusinen, Malenkov, Molotov, Saburov, Chesnokov, candidate Presidium member Yudin, and CC members Pospelov and Rumyantsev (*Pravda*, 9 October 1952).

[72] The Ideological Commission has only been mentioned in occasional obituaries of officials connected with it in 1952–53. A. Ya. Chugayev was identified as a "member of the secretariat of the Ideological Commission under (*pri*) the CC Presidium" from 1952 to 1953 (*Voprosy filosofii*, July 1964, p. 184). L. A. Slepov was also listed as a "member of the secretariat of the Ideological Commission under the CC Presidium" in his obituary in the 27 October 1978 *Moskovskaya pravda*, although no precise dates were given. Khrushchev states that after the 19th Congress Stalin created "among the new Presidium members some wide-ranging commissions" which "in practice" turned out to be ineffectual because no one knew what they were supposed to do (*Khrushchev Remembers*, p. 297). The Ideological Commission was presumably one of these.

[73] Reported in the 28 January 1953 *Pravda*.

[74] No. 2. Another aspect of Chesnokov's role was recently revealed by Aleksandr Nekrich, who was working in the History Institute in early 1953. Commenting on 1953 rumors that Stalin was preparing to exile thousands of Jews to Siberia, Nekrich writes that he knew of a pamphlet written by Chesnokov explaining the need for such a deportation. Nekrich indicates that the pamphlet had already been printed and was awaiting distribution when Stalin died (Nekrich, *Otreshis ot strakha*, p. 114).

[75] *Pravda*, 22 January 1953.

Rumyantsev wrote a 14 November 1952 *Pravda* article on economic laws under socialism and a November 1952 *Voprosy ekonomiki* article on Stalin's theses; and Yudin reported on the theses and attacked other economists and philosophers at a February Academy of Sciences meeting[76] and wrote about Stalin's theses in *Kommunist*, no. 3, 1953.

The new purge campaign was apparently started by a curious attack launched by CC Secretary Suslov in a 24 December 1952 *Pravda* article in which he assailed two articles Fedoseyev had written on the "laws of development of society" (*Izvestiya*, 12 and 21 December 1952).[77] Suslov wrote that Fedoseyev was right in complaining that there were "incorrect, un-Marxist views" among philosophers and economists, but that he had failed to mention that he himself, as chief editor of *Bolshevik* in 1948, had spread such views by printing praise of Voznesenskiy's book. Suslov then revealed that a 13 July 1949 CC decree on *Bolshevik* had criticized the journal for praising Voznesenskiy's book and had specifically attacked Fedoseyev, Aleksandrov, Iovchuk, and Shepilov for backing it. In a 31 December 1952 letter to *Pravda* (published 2 January 1953), Fedoseyev conceded that Suslov's criticisms of his *Izvestiya* articles were correct and confessed error in having praised Voznesenskiy's "anti-Marxist" book in 1948. The fact that some of Suslov's targets were still in positions of authority—Aleksandrov was still Institute of Philosophy director and Shepilov had just been appointed chief editor of *Pravda*—indicated that Suslov's article was an attack on some currently powerful faction, not just beating defeated Zhdanovites. Although most of the targets had been linked with Zhdanov, Malenkov presumably was not behind Suslov's attack, because only months later he was acting as Aleksandrov's patron and Suslov was siding with Khrushchev against Malenkov. Moreover, Shepilov may still have been playing Malenkov's game in 1952, since he reportedly had helped write the CC letter attacking Khrushchev's "agro-town" article in 1951.[78]

Hence, by early 1953 numerous prominent figures in ideology and

---

[76] *Pravda*, 3 February 1953.

[77] A series of major articles on the economic laws of socialism appeared after the publication of Stalin's theses, including: Gladkov in the 10 November *Pravda*, Rumyantsev in the 14 November *Pravda*, and F. Konstantinov in the 11 December *Pravda*.

[78] P. A. Satyukov described Shepilov's role in his speech to the 1961 congress (*Pravda*, 27 October 1961).

economics—Fedoseyev, Aleksandrov, Iovchuk, Shepilov, Anchishkin, Kronrod, Venzher, Notkin, Gladkov, Kozlov, A. Leontyev, L. Leontyev, Gatovskiy, Ostrovityanov—were under attack, suggesting that a shake-up of the whole social science establishment was in the making—a shake-up based on the "truths" revealed in Stalin's new economic theses. These developments suggested an ominous new step backward into dogmatism, as in 1949, or at least a replay on a larger scale of the 1947–48 upheaval in economic science, wherein top economist Varga and others were discredited on the basis of the "truths" revealed in Voznesenskiy's late 1947 book. This time, however, the campaign centered on condemnation of Voznesenskiy's book as erroneous and anti-Marxist and included sharp criticisms of those who had praised it.

Post-Stalin Maneuverings

After Stalin's death on 5 March 1953, his initial successors Malenkov and Beriya quickly adopted relatively "liberal" positions in both domestic and foreign policy, a shift probably encouraged by the fact that they were apparently somewhat dissociated from Stalin's policies at the end—especially his planned purge and crackdown. Beriya repudiated Stalin's campaign against the Jews and his general persecution of non-Russian minorities. Malenkov identified himself with a new campaign to boost consumer goods production and declared that in the nuclear age war was not only not inevitable but virtually unthinkable. However, Khrushchev bested them, in part by implicating them in Stalin's earlier crimes, especially the Leningrad Case. With his dramatic denunciation of Stalin and his rehabilitation of the surviving Zhdanovites, Khrushchev became the patron of moderates, who then served with him against Malenkov.

Beriya began the reversal of Stalin's policies by repudiating the Doctors' Plot, as well as the anti-Semitic campaign and persecution of non-Russian minorities, which were associated with it.[79] At the

[79] A MVD statement in the 4 April 1953 *Pravda* declared the Doctors' Plot a fabrication and a 6 April *Pravda* editorial blamed it on former MGB Minister S. D. Ignatyev and deputy minister M. D. Ryumin. In June Ukrainian First Secretary L. G. Melnikov was fired and publicly condemned for forcing russification on the West Ukraine. Khrushchev later declared that Beriya had submitted memos to the CC Presidium attacking russification in the Ukraine, the Baltic republics, and Belorussia, and that Beriya's memos had caused Melnikov's replacement with a Ukrainian (*Khrushchev Remembers*, pp. 329–30).

suffered in the 1949 purge along with Fedoseyev and some other Zhdanovites, soon allied himself with Khrushchev, delivering one of the key blows in Khrushchev's campaign against Malenkov: as *Pravda* chief editor he authored a long, harsh attack on the consumer goods emphasis, claiming that it was a deviation from the still valid law of giving priority to the development of heavy industry (*Pravda,* 24 January 1955). He was rewarded by Khrushchev by being elected CC secretary in July 1955; however, he lost all his positions when he switched to the anti-Khrushchev side during the "Anti-Party Group's" June 1957 challenge to Khrushchev.[88] P. N. Fedoseyev was named chief editor of the CC's cadres journal *Partiynaya zhizn* when it was reestablished in April 1954, and in 1955 he became director of the Institute of Philosophy.

Those who were most clearly Zhdanovites—Fedoseyev, Iovchuk, and Kedrov—actively contributed to the attacks on Malenkov's faction in 1954–55. The first issue of *Partiynaya zhizn,* with Fedoseyev as editor, editorialized against prominent opponents of Khrushchev's agricultural programs, and the first issue for January 1955 editorially criticized Malenkov's economic policy by assailing an economist's assertion that the "policy of forced rates of development of heavy industry" was outdated. Iovchuk also attacked those who questioned heavy industrial priority in *Voprosy filosofii* (no. 1, 1955, signed to press 8 March 1955). These attacks were clearly connected with Khrushchev's attack, made at the January 1955 CC plenum, on Malenkov's consumer goods program, which Khrushchev labeled in effect a "revival of right deviation" (*Pravda,* 3 February 1955). Fedoseyev followed up more sharply, attacking "some economists" for "trying to prove the need for preferential development of light industry" and terming efforts to end the policy of preference for heavy industry a "very crude distortion of Marxism-Leninism" and a "revival of right opportunism" (*Kommunist,* May 1955, no. 8, p. 33). The January 1955 *Partiynaya zhizn* editorial attacked the "dogmatic approach" in ideology, citing as

---

chief of a department in the foreign ministry 1953–58. (For biographic details, see his biographies in the *Bolshaya sovetskaya entsiklopediya* yearbook for 1962, the 1962 *Deputaty verkhovnogo soveta SSSR,* and the *Filosofskaya entsiklopediya.*)

[88] Apparently judging that the Malenkov-Kaganovich-Molotov group would win because it had the mathematical majority, he voted with them against Khrushchev. The 29 June 1957 CC decree announcing the defeat of the group labeled it the "Anti-Party Group" and defined it as consisting of "Malenkov, Kaganovich, Molotov, and Shepilov, who joined them" (*Pravda,* 4 July 1957).

examples a recent book by conservative philosopher D. I. Chesnokov[89] and another book issued by the Institute of Philosophy. Fedoseyev also wrote a *Partiynaya zhizn* article (December 1955, no. 23) criticizing the institute's journal *Voprosy filosofii* for "scholastic" and "fruitless" articles. In a September 1955 *Kommunist* (no. 14), Kedrov attacked Institute of Philosophy director G. F. Aleksandrov, who was soon replaced by Fedoseyev.

The positions of ex-Zhdanovites like Fedoseyev and Iovchuk appeared guided more by considerations of hostility to Malenkov and concern for personal careers than by consistency or deeply held conviction, and, in fact, they later prospered as relatively conservative guardians against deviations in social sciences. Not unlike Beriya and Malenkov, who switched policies upon Stalin's death, or Khrushchev who changed from a conservative to reform stance, their orientation was determined by circumstances. However, at least one of the old Zhdanov "liberals" did not turn conservative because of old age or out of concern for power and position. B. M. Kedrov took advantage of Khrushchev's destalinization to renew his crusade for reform and internationalism.

Stalin's death set off a remarkable shift to "liberal" positions on the part of his top deputies. First reacting against Stalin's terrorist policies, his successors wound up condemning Stalin himself. The dismantling of Stalin's controls and the later denunciation of Stalin himself opened the door for reformers—such as Kedrov—to begin agitating again, particularly to overthrow the dogmatism imposed in 1948–49.

[89] In late 1952 Chesnokov, who had replaced Kedrov as *Voprosy filosofii* chief editor in 1949, was raised to full membership in Stalin's new CC Presidium, and made co-chief editor of *Bolshevik* and head of a CC section. He lost these posts upon Stalin's death, however (see his obituary in the 18 September 1973 *Pravda*, his biography in the *Filosofskaya entsiklopediya*, and his identification as section head in the 5 February 1953 *Vechernyaya Moskva*).

CHAPTER 6

B. M. Kedrov — A Soviet "Liberal"

The history of moderation in the Soviet establishment does not end with the defeat traced in the preceding chapters. Stalin's death permitted the forces of moderation gradually to recoup, and in time they felt strong enough to attempt to reverse their earlier defeats. The final dethroning of Lysenko in 1964–65 provided an especially good opportunity to strike back at dogmatism and even to challenge party restraints on social and natural sciences. This chapter traces the fortunes of moderation in 1946–49 and later through a close study of the career of venerable Soviet philosopher Bonifatiy Mikhaylovich Kedrov.

To study Kedrov is also to address the basic question of whether "liberals" or "moderates" can really exist in the Soviet "establishment." Previous chapters demonstrate the existence of a variety of political viewpoints in the Soviet leadership, even during the harsh postwar period, and Kedrov's story provides further evidence that it is possible for independent thinkers to exist within the Soviet political establishment, despite the system's overwhelming stress on conformity and ideological orthodoxy. His career in influential positions spans over thirty years, during which he has been repeatedly ousted for his unorthodox views. His demotions hardly evoke surprise, but his ability to return to prominence and his persistence in again taking controversial stands are remarkable and suggest that others in the party, government, and academic apparats sympathize with his aims, even while not expressing them as openly. Naturally, most unorthodox thinkers dare not flaunt their views if they wish to retain their positions and privileges—a situation that makes it difficult to obtain hard evidence of their "liberalism." In Kedrov's case, the evidence is clear and abundant.

Kedrov has figured prominently in three contests between "liberals" and "dogmatists"—the 1947–49 struggle over internationalism in science and philosophy, the 1965–68 struggle to liberate science and philosophy from Lysenkoism, and the 1973–74 struggle to emancipate philosophy and other social sciences from doctrinaire ideological constraints. He has sought to replace the crude dogmatism that usually dominates Soviet social sciences with a kind of enlightened communism—a communism that could prevail in the struggle of ideas by superior arguments rather than by force or name-calling. He argues that Western scientific concepts should be accepted or debated on their merits, not simply rejected on the grounds that they are capitalist, bourgeois, or non-Russian. Lastly, Kedrov has campaigned to obtain for social scientists the same freedom to pursue new ideas without narrow ideological restraints that Soviet natural scientists possess.

Kedrov's ideas have come into conflict with basic trends in Soviet foreign policy. The defeat of his ideas in 1947–49 accompanied the development of the cold war and Soviet Russian rejection of foreign ideas and influence. The defeat of his views in 1973–74 was an aspect of the debate over the basic meaning of detente: whether relaxation of tensions abroad was to be accompanied by any relaxation of controls at home. Soviet leaders maintained that a soft line internationally (détente) must not lead to ideological erosion at home and must therefore be accompanied by a hard line domestically, whereas Kedrov and his allies had argued unsuccessfully that détente required liberalization at home.

In view of Kedrov's repeated defeats, one may ask whether his efforts have had any effect. They clearly did in 1947–48—until Stalin himself stepped in to endorse the forces of dogmatism and remove Zhdanov, who had afforded some protection to relatively enlightened thought in science and philosophy. The 1965–68 campaign against Lysenko led to the overthrow of the dogmatists controlling philosophy. And, most recently, Kedrov's daring 1973–74 campaign to liberalize Soviet social sciences at least created significant ripples in the academic community and was overcome only with difficulty by the dogmatists. Though the long-term effect of Kedrov's efforts remains unclear, his ideas have clearly influenced other social scientists and are likely to reemerge in the future.

Start of an Improbable Career

Kedrov's original rise to prominence was improbable. A man whose father and brother had been executed as "enemies of the people," he was appointed by Zhdanov, the supposed archdogmatist, to leading posts that involved the enunciation and interpretation of Marxist philosophy during the postwar ideological crackdown (the *Zhdanovshchina*). His presumably precarious position did not inhibit him from striking out boldly and presenting controversial ideas.

Kedrov's father, Mikhail Sergeyevich Kedrov, had been a prominent Old Bolshevik, a friend of Lenin,[1] and a high official of the new Soviet government.[2] In the first years after the revolution, he was, in fact, a top leader of the secret police (*Cheka*) and had a reputation for ruthlessness toward foes of the Bolsheviks. During most of the 1920s and 1930s, however, he was not involved in police work.

As a result of assignments in Baku in the early 1920s, the elder Kedrov apparently learned compromising facts about the young *chekist* L. P. Beriya, then active in that region. According to an article in *Leningradskaya pravda* (25 February 1964), Kedrov, after reviewing the work of the Azerbaydzhan *Cheka* in 1921, had urged that Beriya be fired from the *Cheka*. In February–March 1939, after Beriya's appointment as USSR NKVD chief, Kedrov and his youngest son Igor "spoke out against the hostile activities of Beriya," sending Stalin several letters revealing cases of traitorous activity by Beriya. "As a result of this, Igor Kedrov was arrested and then shot," and Kedrov senior was arrested, according to this article. Although acquitted of wrongdoing by the military collegium of the Supreme Court, Kedrov senior was executed anyhow on the order of Beriya, as Khrushchev indignantly explained in his 1956 secret speech. *Pravda* (17 December 1953), in detailing the case against Beriya,

[1] Kedrov senior went into exile in Switzerland in 1912 where he became a friend of Lenin (*Izvestiya*, 11 July 1970, and *Pravda*, 24 February 1978). Kedrov junior, in a December 1973 *Voprosy filosofii* interview, recalled Lenin's frequent visits.

[2] See his biographies in the *Bolshaya sovetskaya entsiklopediya* (3d edition) and *Sovetskaya istoricheskaya entsiklopediya*, as well as details in the 25 February 1964 *Leningradskaya pravda*.

stated that he had killed Kedrov because he "suspected that Kedrov had material on Beriya's criminal past." Thus, young B. M. Kedrov clearly had a dangerous enemy—secret police chief Beriya—as he began his career.

Kedrov had spent his early career studying chemistry and only switched to philosophy in 1931, when tuberculosis forced him to give up chemical lab work. After devoting the 1930s to scientific work, he joined the army in 1941 and was demobilized only after the war ended.[3]

Immediately upon leaving the army in 1945, Kedrov was appointed deputy director of the Institute of Philosophy—an appointment that demonstrates his good patronage ties with someone. How he was chosen is not clear, but those who may have had a voice in this appointment presumably included the ideological overseer Zhdanov, Agitprop chief Aleksandrov, Academy of Sciences President S. I. Vavilov (brother of Lysenko's arch-rival in biology, N. I. Vavilov), and Institute of Philosophy director V. I. Svetlov. All of these men later appeared to have good relations with Kedrov: Zhdanov accepted Kedrov's 1947 proposal for creating a philosophy journal and named him chief editor; Vavilov wrote the introduction to a 1949 Kedrov book;[4] Svetlov and his successor G. S. Vasetskiy[5] sided with Kedrov in the June 1947 philosophy debate. Although relations between Aleksandrov and Kedrov later deteriorated, they were very good in 1946–47: Kedrov, as deputy director, got Aleksandrov's *Istoriya zapadnoyevropeyskoy filosofii* approved for a Stalin Prize by the Institute of Philosophy over the opposition of S. L. Rubinshteyn, the first deputy director.[6] In the June 1947 philosophy debate, Kedrov appeared to try to divert criticism from Aleksandrov to the conservatives Mitin and Yudin.

Champion of Internationalism, 1947–49

During 1947–49 Kedrov became the most outspoken champion of "internationalism" in science and philosophy and the leading

---

[3] The details of Kedrov's biography can be gleaned from his December 1973 *Voprosy filosofii* interview and from the entry on him in the *Bolshaya sovetskaya entsiklopediya*, 3d ed.

[4] See *Voprosy filosofii*, December 1973, p. 160.

[5] In 1946 Svetlov left to become deputy minister for higher education, according to his biography in the *Filosofskaya entsiklopediya* (1970). Vasetskiy's appointment was reported in the July 1946 *Vestnik* of the Academy of Sciences, p. 81.

[6] According to Maksimov in the June 1947 philosophy debate (*Voprosy filosofii*, no. 1, 1947, p. 190).

target of dogmatists and chauvinists. He argued the importance of objective scientific knowledge and insisted that Soviet science should not cut itself off from international scientific achievements. He contended that his was the true Marxist position, putting a class, internationalist approach over narrow national viewpoints. However, his position was overwhelmed by the rising tide of Russian chauvinism and the "two camps" theory, and Kedrov was pilloried as "cosmopolitan" and unpatriotic.

Kedrov's role in promoting Aleksandrov's book and in putting forward controversial ideas had guaranteed him a place in the center of the 1947 philosophy debate, and his sharp comments drew even more fire. One of the first to speak at the June 1947 meeting, Kedrov contended that the Aleksandrov book included much that was valuable, as well as bad points, and that at least Aleksandrov had tackled questions of history of philosophy instead of just compiling quotes from Marx. Then Kedrov tried to turn the debate against Mitin and Yudin by recalling the errors of their 1944 history of philosophy. Mitin struck back by accusing Kedrov of trying to protect Aleksandrov's book[7] and by asserting that Kedrov's early 1947 book *O kolichestvennykh i kachestvennykh izmeneniyakh v prirode* made similar (vague) errors and reflected "menshevizing idealism."[8] Kedrov's friends came to his defense: P. Ye. Vyshinskiy rebuked Mitin for calling Kedrov a "menshevizing idealist"[9] and I. A. Kryvelev declared that Kedrov was "a working and creative person," whereas Mitin only "chews over old quotations."[10] Despite the fact that most criticism of the Aleksandrov book accused it of underrating Russian philosophy, Kedrov in the June 1947 debate chose to interpret the book's errors as precisely the opposite: exaggerating the role of Russian philosophy and underrating the international aspect of philosophy. Although some speakers (V. P. Chertkov and V. S. Molodtsov) criticized Kedrov's comments for downplaying the importance of national (i.e., Russian) philosophy, Zhdanov and other key figures did not.

Kedrov also used the June 1947 debate to propose to Zhdanov the creation of a special philosophy journal and soon after the session, Zhdanov sanctioned the creation of *Voprosy filosofii* and Kedrov

[7] Ibid., p. 122.
[8] Ibid., p. 127.
[9] Ibid., p. 231.
[10] Ibid., p. 393.

was named chief editor. Kedrov immediately began using the new journal to promote liberal ideas in philosophy, physics, biology, and other fields,[11] and to fight such dogmatists as Lysenko and Maksimov. However, he only succeeded in putting out three issues (nos. 1 and 2, 1947, and no. 1, 1948) before Zhdanov died and Kedrov lost control of the journal.

With his provocative viewpoints, Kedrov quickly became the target of conservative attacks. A. A. Maksimov, the conservative physicist, attacked Kedrov's May 1947 book, *Engels i yestestvoznaniye* in the 31 December 1947 *Kultura i zhizn* and repeated his attacks in a January discussion of the book held at the Institute of Philosophy. However, while other speakers (such as Aleksandrov) did acknowledge that Kedrov had overstressed the international character of science, Maksimov's attacks found no substantial support at the meeting.[12] And Kedrov, at an April meeting in the Academy of Sciences, lashed back, accusing Maksimov of emphasizing Russian science over Marxism.[13]

As Zhdanov fell into disfavor, however, attacks on Kedrov increased and soon prevailed. On 17 July 1948, a *Literaturnaya gazeta* editorial attacked Kedrov's book on Engels for downgrading the importance of Russian science and taking a "nihilistic attitude to questions of national culture." Calling his views "cosmopolitan," the editorial condemned the idea of an abstract world science. *Voprosy filosofii*, no. 2, 1948 (signed to press in October 1948), still listed Kedrov as chief editor, but editorialized against his book on Engels and assailed opposition to Lysenko. Shortly thereafter, Kedrov was removed as chief editor and *Voprosy filosofii* joined other publications in attacking him as anti-Lysenkoist, unpatriotic, and cosmopolitan. For example, in *Voprosy filosofii*, no. 3, 1948 (actually signed to press only in June 1949), Iovchuk criticized Kedrov for a "cosmopolitan" approach in denying the importance of Russian national philosophy. In a 9 March 1949 *Literaturnaya gazeta*, Mitin called Kedrov the "ideological inspirer of the group of persons spreading cosmopolitan views," and said that he was "nihilistic" toward "Russian national culture." Mitin claimed that Kedrov was pushing the idea of "a world science that knows no national or state

[11] For example, printing controversial articles by M. A. Markov on physics and I. I. Shmalgauzen on biology in *Voprosy filosofii*, no. 2, 1947.

[12] See the account in the March 1948 *Vestnik* of the Academy of Sciences.

[13] See the account in the June 1948 *Vestnik* of the Academy of Sciences.

borders." The 10 March 1949 *Kultura i zhizn* quoted Kedrov as having said that "there is a single, world science" and that it is un-Marxist to consider philosophy country by country.

Thus, the long debate between Kedrov and the conservatives over internationalism in science and philosophy ended with the victory of those who insisted that science and philosophy were divided into capitalist and communist, and that Russian national science and philosophy must be stressed. In 1949 Kedrov was forced to confess error in having fought against Lysenkoism and the notion of Russian priority in the social and natural sciences. After a 10 March 1949 *Kultura i zhizn* editorial article assailed Kedrov for downgrading Russian philosophy and science, Kedrov confessed that he had indeed neglected the priority of Russian science and scorned Lysenko (letter printed in *Kultura i zhizn*, 22 March 1949).

Kedrov was dropped from the Institute of Philosophy in 1949, as well as from the post of *Voprosy filosofii* chief editor, and did little meaningful work until 1953. His new book, *Mirovaya nauka i Mendeleyev* (World science and Mendeleyev), was pulled back from the publishing house in 1949 and canceled, a fact later noted by Kedrov in his December 1973 *Voprosy filosofii* interview (p. 160). The 10 March 1949 *Kultura i zhizn* accused Kedrov of having portrayed Mendeleyev as a supporter of "world science" in his earlier book on Engels and natural science. Kedrov remained a member of *Voprosy filosofii*'s editorial board and a professor at the Academy of Social Sciences, and also went to work on the *Bolshaya sovetskaya entsiklopediya* during 1949–52.[14] After 1949, he was forced to write pro-Lysenko articles. As Ye. Kh. Frauchi later stated in a letter to the journal *Oktyabr* defending Kedrov, as "son and brother of 'enemies of the people,'" Kedrov had lived under a "Damoclean sword of possible repression" and "had to speak out against his own true convictions."[15]

Kedrov began a slow comeback after Stalin's death. During the Malenkov-Khrushchev struggle, he served Khrushchev's side by writing a sharp attack on Malenkov's minister of culture, G. F. Aleksandrov, in a September 1955 issue of *Bolshevik* (no. 14).[16] In 1958 he rejoined the Institute of Philosophy, in 1960 he was elected a

---

[14] See his biography in the third edition of the *Bolshaya sovetskaya entsiklopediya*.
[15] *Oktyabr*, February 1966, pp. 144–45.
[16] Aleksandrov had developed ties to Malenkov, who appointed him minister of culture in early 1954. Following Malenkov's resignation as premier in early 1955,

corresponding member of the Academy of Sciences, and in 1966 he was elected a full member. In 1962 he became director of the Institute for the History of Natural Science and Technology, the base from which he would again rise to prominence.

Leader in Anti-Lysenko Drive, 1965–68

Even as leader of his new, seemingly innocuous institute, Kedrov returned to controversy in the mid-1960s, joining the attack on Lysenkoism and striving to reestablish the internationalism of science and to expand the freedom of scientific inquiry. With the October 1964 fall of Khrushchev, who had become Lysenko's main patron, Kedrov and others quickly produced articles aimed at discrediting Lysenko and ending his reign in science. Moreover, the drive to free biology from dogmatism was apparently seen by Kedrov as the first step to the liberation of other natural sciences and eventually the social sciences as well.

The January 1965 *Novyy mir* carried a remarkably frank article by Kedrov, in which he sharply attacked dogmatism in biology and other sciences. Welcoming the post-Khrushchev revelations of Lysenko's abuses, he assailed Lysenko's "persecution" of foes in science and demanded that Lysenko's practice of using labels like "Weismannism-Morganism-Mendelism" be renounced "once and for all." He recalled that it had been possible to "get rid" of persons simply by labeling them "enemies of the people" and that the "patriotism of Russian natural scientists" had been equated exclusively with Lysenkoism. He complained that some people misunderstood *partiynost* (party spirit), seeing in it simply the need to hunt for "idealism" in new natural science theory and "to curse it in terms as strong as one likes." Such a conception of *partiynost,* he declared, boiled down to passing "personal opinions on new scientific theories," and thus blocked any new ideas that conflicted with accepted theories. He recalled how A. A. Maksimov had fought Einstein's theory of relativity in the 1940s and 1950s and how cybernetics had been declared a "bourgeois pseudoscience." Kedrov argued that Soviet science should use such Western scientific

---

Aleksandrov lost the post. In the December 1973 *Voprosy filosofii* interview, Kedrov recalled his 1955 attack on Aleksandrov, who, he declared, during 1953–54 had headed a group of philosophers who incorrectly interpreted dialectics.

achievements, once it has stripped them of their incorrect bourgeois ideological aspects, but that "vulgarizers like A. A. Maksimov" preferred to throw out not only ideologically objectionable features but the entire scientific idea. Kedrov also helped prod educational authorities to remove Lysenkoist teachings from school biology texts, attending 1965 meetings of a commission to prepare new biology curricula.[17]

In the November 1966 and May 1967 issues of *Molodoy kommunist,* Kedrov argued that dialectical materialism and natural sciences were compatible, despite the damage that Lysenkoism had inflicted on Soviet science. He found a valuable ally in the young philosopher I. T. Frolov, who wrote articles in the 15 March 1967 *Pravda* and January and August 1967 issues of *Voprosy filosofii,* in which he attacked Lysenkoism for tending to discredit dialectics. In his March *Pravda* article, Frolov noted that the citation of dialectical materialism to condemn many new ideas in natural science had "discredited [dialectics] in the eyes of scholars" and given grounds to those foes who claimed that dialectical materialism was an unsuitable method of natural science. "There can be no return to those times when Lenin's idea of uniting philosophers and natural scientists was interpreted formally and superficially and was undermined by incompetent interference in matters of science, often leading to discrediting the very idea," he wrote. In the January *Voprosy filosofii* article, Frolov recalled that Lysenkoist philosophers had used materialist dialectics to condemn the theory of genes, but insisted that materialist dialectics has nothing in common with the "vulgar form" given it by Lysenko's defenders. Frolov also wrote a 1968 book, *Genetika i dialektika* (Genetics and dialectics), attacking Lysenko and arguing for more independence for scientific investigation.

Kedrov quickly became the target of conservatives again. In the August 1965 *Oktyabr,* the Lysenkoist G. V. Platonov concentrated on attacking Kedrov for his criticisms of Lysenko, and in February 1966 the conservative editors of *Oktyabr* printed comments by Lysenkoist apologists who accused Kedrov of striving to drive *partiynost* out of natural sciences and to accept Western science without regard for ideology. V. I. Mineyev of the Far East State University complained that Kedrov is "striving to separate scientists' ideological positions from 'their real contribution to the develop-

---

[17] See I. I. Gunar in *Uchitelskaya gazeta,* 3 June 1965.

ment of science'" and is "calling on us to close our eyes to the connection between a natural scientist's scientific views and his ideological position."[18] Kiev philosopher T. K. Martynenko attacked Kedrov's January 1965 *Novyy mir* article for urging that Western scientists be evaluated only as scientists, regardless of their political views, and accused Kedrov of "seeking in the works of bourgeois scholars only that which is acceptable, ignoring that which could be not only unacceptable but even extremely harmful for our youth."[19]

Frolov was assailed in similar terms. In an article in *Selskaya zhizn* (17 May 1969), agricultural scientist V. Shaykin criticized Frolov's 1968 book on genetics and dialectics for one-sidedly attacking Lysenko and for seeking to separate science from ideology. Shaykin accused Frolov of denying that philosophy and the natural sciences should be just as party-oriented as the economic and historical sciences, and he insisted that genetics "cannot stand outside the struggle of world views."

Lysenko and his main defenders in biology lost their power in 1965, but attacks on Lysenko were largely halted in late 1965, apparently to prevent the campaign from damaging the reputations of various powerful party figures and undermining party control over science. Nevertheless, the upheaval in biology shook the control that the dogmatists, who were inextricably entangled with Lysenkoism, exercised over philosophy, a process that took about three years to complete, however.

In early 1968, the conservatives suddenly lost control of the Institute of Philosophy and its organ, *Voprosy filosofii*. At a general meeting of the Academy of Sciences (6–7 March 1968), P. V. Kopnin, the 46-year-old director of the Ukrainian Institute of Philosophy, was elected to replace F. V. Konstantinov, a 67-year-old ultraconservative, as director of the USSR Institute of Philosophy, and in May or June 1968, 39-year-old I. T. Frolov replaced the 67-year-old Lysenko ally M. B. Mitin as chief editor of *Voprosy filosofii*.[20]

Since Kopnin died in June 1971,[21] he had relatively little opportu-

[18] *Oktyabr*, no. 2, 1966, p. 148.
[19] Ibid., p. 149.
[20] For Kopnin's election, see the May 1968 *Vestnik akademii nauk SSSR*, p. 140. Frolov first signed as chief editor in the June 1968 *Voprosy filosofii*, signed to press on 19 June.
[21] See the 29 June 1971 *Pravda* for his obituary.

nity to influence philosophy, but had apparently already come into conflict with the dogmatists. In a 7 March 1975 Radio Liberty interview,[22] émigré sociologist Ilya Zemtsov described how Kopnin, dejected over KGB-ordered persecution by other philosophers, declared on his deathbed that he did not want to live since he did not want to bow to "scoundrels." (This statement was made during the 1970–71 attacks on liberal sociologists.)

Frolov, who soon turned out to be the most controversial of the reform-minded Soviet philosophers (other than Kedrov himself), quickly embarked on a program to stimulate debate and promote new ideas in philosophy. At the end of February 1969, the Presidium of the Academy of Sciences discussed a new program for the journal which had been prepared by Frolov. In his speech, Frolov stressed that he saw the journal's "main function" as raising sharp questions and declared that its "main section" must be the "debate section." He also planned more effective criticism of bourgeois ideology and a broader international distribution of the journal. (Frolov had written in the 15 March 1967 *Pravda* article that "open competition with hostile views" and "encounters with the ideological opponent on his own 'territory'" were not to be feared.) The *Voprosy filosofii* account of the meeting (May 1969) reported enthusiastic response by academy leaders. P. L. Kapitsa, the independently minded physicist, expressed particular praise for Frolov's plan to expand work in the international competition of ideas. Declaring that "our ideologues" should compete internationally, Kapitsa stressed that they must first improve their abilities, since they will "lose the privilege that they have in our country, where they encounter no opposing views." Academy Vice President M. D. Millionshchikov and P. N. Fedoseyev, the director of the Institute of Marxism-Leninism, seconded Kapitsa on this point.

In a sense, Frolov's promotion to chief editor of *Voprosy filosofii* in 1968 is easy to understand, for he had links with two leaders who were very influential at this time: P. N. Demichev, the ideology secretary of the CC and Politburo candidate member, and A. M. Rumyantsev, vice-president for social sciences of the Academy of Sciences. In 1963 Rumyantsev, then chief editor of *Problemy mira i sotsializma,* had hired the thirty-four-year-old Frolov as deputy to

[22] In Radio Liberty's Russian-language analysis series no. 106 for 1975 (RS 106/75).

the responsible secretary of his journal.[23] By late 1965, Frolov had become personal assistant to Demichev, the CC ideology chief.[24] Despite the general ban on articles attacking Lysenko after 1965, Demichev permitted his assistant to write three articles in 1967 criticizing Lysenko and urging renovation in philosophy, after which Frolov was appointed to the key post of *Voprosy filosofii* editor. The Frolov and Kopnin appointments and the ouster of conservative philosophers presumably also stemmed from the work of A. M. Rumyantsev, the liberal who had become vice-president of the Academy of Sciences and head of the academy's social science section in 1967. Rumyantsev also lobbied successfully for creation of a sociology institute (the Institute for Concrete Social Research [*Institut konkretnykh sotsialnykh issledovaniy*]), of which he became director in 1968—another important step in liberalizing and invigorating Soviet social sciences.

Sharp attacks on liberal social scientists, especially sociologists, increased in 1970, however, and by 1971 conservatives had regained the upper hand.[25] After sharp infighting, Rumyantsev was removed as vice-president of the Academy of Sciences and as director of the Institute for Concrete Social Research in 1971. M. N. Rutkevich, a

[23] *Zhurnalist*, July 1968, p. 79.

[24] A. Solzhenitsyn met Frolov on 21 October 1965 during a visit to Demichev and gave his title as *referent* (consultant) on questions of culture for Demichev. See Solzhenitsyn's *Bodalsya telenok s dubom* [A calf butts against an oak] (Paris, 1975), pp. 131–32. (For the English version of this passage, see the translation of the book published as *The Oak and the Calf* [N.Y., 1979], p. 116). Aleksandr Nekrich mentions talking with Frolov in October 1966, at which time the latter was *pomoshchnik* (assistant) to Demichev (Nekrich, *Otreshis ot strakha*, p. 265). Frolov was also identified as Demichev's aid prior to 1968 in émigré sociologist Ilya Zemtsov's book *Partiya ili mafiya?* [The Party or the Mafia?] (Paris, 1976), p. 38. He was identified in the 1 November 1966 *Pravda* as a "CPSU CC official," accompanying Demichev and Agitprop head V. I. Stepakov on a trip. He was again listed as a CC official in the 3 March 1967 *Moskovskaya pravda*.

[25] Attacks on the new sociology institute had begun immediately after its creation. For example, in the 3 August 1968 *Komsomolskaya pravda*, M. N. Rutkevich warned against attempts to create a sociology independent of Marxist philosophy. Others joining this campaign against liberal-minded sociologists included G. Ye. Glezerman, the deputy rector of the Academy of Social Sciences, V. N. Pilipenko, the head of the CC's philosophy sector, and P. N. Fedoseyev, director of the Institute of Marxism-Leninism. Conservatives Glezerman and Pilipenko warned in a March 1971 *Kommunist* (no. 4, p. 67) that creating a sociology separate from historical materialism "opens channels for penetration of bourgeois concepts," while Fedoseyev in a January 1972 *Kommunist* (no. 1, p. 75) criticized sociologists for "copying schemes and models originated in bourgeois sociology" and "underrating" the importance of historical materialism. The campaign ended with the reassertion of tighter ideological controls over sociology, which began to loosen only several years later.

conservative, took over the institute (which was then renamed the Institute for Sociological Research) and purged its liberal officials, while Institute of Marxism-Leninism director P. N. Fedoseyev became academy vice-president and lined up with the conservatives. The loss of their liberal patron Rumyantsev was a setback for the liberals in philosophy, and when the Society of Philosophers was founded in December 1971, the conservative F. V. Konstantinov became president.[26] When Kopnin died in June 1971, a stalemate apparently developed over naming a successor, and the Institute of Philosophy thus proceeded without a director for almost two years.[27]

Champion of Liberalization in Social Sciences, 1973–74

The second period of Kedrov's influence on high-level issues began when he was named director of the Institute of Philosophy in early 1973. Immediately, he embarked on an ambitious plan to liberalize and invigorate philosophy and social sciences in general. The campaign was short-lived, however, for it alarmed the party watchdogs and even challenged the leadership's plan to couple détente abroad with ideological vigilance at home. Within a year Kedrov was forced to resign from the Institute of Philosophy, and a long campaign to undo his influence began.

Kedrov's second rise is just as difficult to explain as his first. Liberals were certainly not in control, and the philosophy community was so deeply divided that the institute had gone without a director for two years. The leadership clearly knew Kedrov's views, and many of those involved in supervising philosophy would have opposed him strongly, notably the ultraconservatives who headed the CC Science Section (S. P. Trapeznikov) and the Academy of Sciences philosophy and law division (F. V. Konstantinov). In fact, Trapeznikov began feuding with Kedrov and Frolov almost immediately. Moreover, Suslov, the Politburo's ideological overseer, would hardly be any more sympathetic to Kedrov in 1973 than he

[26] See the January 1972 *Vestnik akademii nauk SSSR*, p. 132, for Konstantinov's election.

[27] The split in philosophy can be seen in the membership of the philosophy and law division of the Academy of Sciences in late 1971. Headed by academic secretary Konstantinov, the division's five full members included both the conservatives Mitin and Konstantinov and the liberal Kedrov (the other two were Fedoseyev and L. F. Ilichev), while the corresponding members included the conservatives Maksimov and Rutkevich and the more moderate D. M. Gvishiani and Iovchuk.

had been in 1947–49. In view of Kedrov's removal after only a few months, it is clear that whoever sanctioned his appointment had made a serious mistake. It is probably not accidental that Frolov's former chief and patron, P. N. Demichev, was removed as CC ideology secretary shortly after the crackdown on Frolov and Kedrov began.

Kedrov became director of the Institute of Philosophy in March or April 1973[28] and announced his intentions to reform philosophy in an article in the September 1973 issue of *Voprosy filosofii.* Here he called for a decisive turn toward creativeness in philosophy and condemned the crude methods used by conservatives to silence their foes and restrict scholarly inquiry. He declared that the Institute of Philosophy must lead philosophy and "really move philosophy forward" rather than just "repeat long-known truths." At the same time, he demanded that more effective, intellectually respectable methods be used in refuting anti-Marxist philosophy. Rather than resorting to rude name-calling and attacks on phrases removed from their context—tactics typical of Soviet polemics—Kedrov insisted that "serious criticism must be based on deep study of the work being criticized." Moreover, he declared that if such crude methods were wrong to use on Western foes, they were even more objectionable when directed at fellow Soviet philosophers ("if such 'criticism' is impermissible regarding our ideological opponents, it is impermissible also regarding our own philosophical works"). In an apparent attempt to reduce the ideological restrictions that hampered innovative research in philosophy and the social sciences, Kedrov said that philosophy and other social sciences should be brought closer to the natural sciences. Since 1962, as director of the Institute for History of Natural Science and Technology, Kedrov had actively fostered the rapprochement of these fields.

Kedrov's campaign was also pressed by his close ally Frolov, chief editor of *Voprosy filosofii,* and some other social scientists. On 25 April 1973, *Voprosy filosofii* and the Academy of Social Sciences (under rector M. T. Iovchuk) organized a "round table" on cooperation between the natural and the social sciences.[29] During this

---

[28] He was first identified as director of the Institute of Philosophy at a 25 April 1973 conference reported in the September 1973 *Voprosy filosofii,* p. 42. At a 14–15 March 1973 conference reported in the June 1973 *Voprosy filosofii,* p. 39, he was still listed as director of the Institute for History of Natural Science and Technology.

[29] *Voprosy filosofii,* nos. 9 and 10, 1973.

discussion, Frolov outlined an ambitious goal—that philosophers should become the leaders (*zastrelshchiki*) in promoting a much closer exchange between these two fields of study.[30] The apparent political purpose of this campaign was revealed most clearly by Professor Kh. N. Momdzhyan (head of the department of philosophy at the Adacemy of Social Sciences), who declared that closer cooperation meant that the social scientist must be just as free to pursue objective knowledge as his colleague in the natural sciences: "It cannot be that in one science people work on discovering new scientific truths, while in a number of branches of social sciences people have been busy repeating the same old propositions for decades." Complaining of the lack of creativity in philosophy, historiography, and economics, Momdzhyan insisted that the social sciences must be as innovative as the natural sciences. Moreover, he warned against crude assaults by conservatives and party watchdogs, stating that "none of us intends to abandon party positions in theory," and "if anyone thinks that his task in science is only to take a club and hit on the head anyone who expresses his ideas differently, nothing useful will result." It is necessary to "learn to speak a language accepted among scholars," he continued.[31] Sounding one of Kedrov's favorite themes, Iovchuk criticized the "mistaken" idea that science is not universal.[32] Frolov, Kedrov, and Iovchuk enlisted a number of prominent natural scientists to join their "round table" and thus lend support to the philosophers' campaign.[33] Kedrov concluded the discussion by stating that Marxism-Leninism is "not a collection of ready-made formulas but creative, living doctrine, presupposing study and search for new aspects of the subject under study."[34]

These points were repeated forcefully by Frolov in a special edito-

---

[30] Ibid., no. 9, 1973, p. 59.

[31] Ibid., no. 10, 1973, pp. 48–49.

[32] Iovchuk appeared sympathetic to Kedrov's campaign. Iovchuk had hired Kedrov as a professor at the Academy of Social Sciences in 1971 shortly after becoming rector (see Kedrov's biography in the third edition of the *Bolshaya sovetskaya entsiklopediya*) and in the April 1973 discussion praised Kedrov's work as director of the Institute for History of Natural Science and Technology.

[33] From reform-minded mathematical economists like N. P. Fedorenko and L. V. Kantorovich to anti-Lysenko biologist N. P. Dubinin. Fedorenko wrote a separate article in the October 1973 *Voprosy filosofii* on ties between the natural and social sciences, attacking foreign philosophers for attempting to use achievements in biology, math, and physics to discredit dialectical materialism.

[34] *Voprosy filosofii*, no. 10, 1973, p. 53.

rial on *partiynost* in the January 1974 *Voprosy filosofii*. While admitting defects in *Voprosy filosofii*'s work, the editorial turned on those whose role in philosophy is not to create but to sniff out and condemn any idea that deviates from orthodoxy. This is phony *partiynost*, the editorial declared; real *partiynost* presupposes "objectivity of philosophic knowledge" and pursuit of "truth no matter how bitter and mercilessly critical it may be,"[35] not "cavalry raids" (*kavaleriyskiye naskoki*) on scholars who seek to expand knowledge.

Unfortunately, we do not always notice that it is the people who are dissatisfied with everything and criticize everyone, while themselves doing nothing or extremely little for our philosophy, who try to create a halo of infallibility around themselves. Sometimes lack of talent and professional skill dresses itself in the toga of "ultra-*partiynost*" and takes refuge in loud phrases.

Fake *partiynost*, intended to mask scientific fruitlessness, has an especially negative effect on discussion of new problems and organization of debate over particular books and articles on philosophy. Here subjectivism and group interests often prevail over genuine scientific and party interests....

Scientific criticism, from positions of Marxist-Leninist *partiynost*, has nothing in common with the "cavalry raids" still occurring among us, during the course of which methods used to fight bourgeois ideology and anticommunism are sometimes transferred to debate among like-minded Marxist philosophers. Instead of businesslike, quiet discussion aimed at improving work and at helping those who have erred, ideological labels insulting for Soviet scholars are hung on people.

Yet it was V. I. Lenin who said: "He who does nothing makes no mistakes." Of course, there are different kinds of mistakes. But if we aspire to explore and do not abandon a Marxist-Leninist outlook, we can come to dissimilar results. Therefore, we need debate, criticism, and self-criticism. But there cannot be in philosophy a kind of "division of labor" whereby some work and try to study new problems, while others only specialize in criticism and themselves do not participate in creative work.[36]

Frolov also brought in détente to use as an argument for liberalization. He recalled Brezhnev's August 1973 Alma-Ata statement (rarely quoted in the Soviet Union) to the effect that détente involved

[35] Ibid., January 1974, p. 47.
[36] Ibid., January 1974, p. 55.

the widening of contacts with the West, a process that both opened new opportunities to spread communist ideas and forced Soviet ideological workers to work more effectively to counter Western influence at home. This demonstrated, according to the editorial, that détente thus necessitated more sophisticated, convincing philosophical writings, clearly implying the need for a philosophy that could compete in the struggle of ideas intellectually, not by name-calling and the censorship of opposing views. This problem was raised again by Kedrov in his September 1973 *Voprosy filosofii* article, in which he called on Soviet philosophers to develop a scholarly refutation of French biologist Jacques Monod's attempt to use Lysenkoism to prove that Soviet Marxism was incompatible with true science. Frolov himself undertook this task, writing an entire article refuting Monod in a February 1974 *Kommunist* (no. 3).

Kedrov, Frolov, and Momdzhyan were clearly aiming their talk of "cavalry raids" and the obstruction of innovative thought at Trapeznikov, head of the CC Science Section, who was well known for his pro-Stalin versions of history and his vicious attacks on any unorthodox ideas. In late May 1973, shortly after Kedrov's appointment as director of the Institute of Philosophy, Trapeznikov had called a conference of philosophers in his CC section to warn Kedrov and Frolov not to press innovation. Trapeznikov's speech at the conference, published as an article in the August 1973 *Voprosy filosofii,* assailed the "conviction among philosophers" that "fancied innovation" (*mnimaya novizna*) is a "means of creative development of Marxist-Leninist philosophy." He asserted that Western foes are using these deviations from *partiynost* in social sciences to harm the USSR. Instead of marking an end to the war of ideologies, he stated, détente demands that we intensify our uncompromising ideological struggle, and philosophers, as "the vanguard of all Soviet social science," must be more aggressive in this struggle.[37] However, as Kedrov's September 1973 *Voprosy filosofii* article and Frolov's January 1974 *Voprosy filosofii* editorial showed, these two innovators had no intention of heeding Trapeznikov's warning.

The conflict virtually became open at a 5 February 1974 meeting

[37] A toned down version of the May discussion appeared in the August 1973 *Voprosy filosofii,* along with Trapeznikov's article. *Voprosy filosofii* was criticized at the May conference, according to the January 1974 *Voprosy filosofii* special editorial (p. 50), although this was not mentioned in the August 1973 account of the conference.

of the Institute of Philosophy, called to examine the work of *Voprosy filosofii*. The institute's deputy director, F. T. Arkhiptsev, who headed the commission named to report on *Voprosy filosofii*'s work, reflected the pressure from Trapeznikov and the head of the academy's philosophy division, Konstantinov. While praising *Voprosy filosofii*, Arkhiptsev acknowledged that it had not carried out all the tasks set in Trapeznikov's August 1973 *Voprosy filosofii* article, and cited the "alarm" over *Voprosy filosofii*'s analysis of dialectical materialism—an alarm "expressed during all of 1973," both in the CC Science Section and by leaders of philosophy in the Academy of Sciences. According to Arkhiptsev, critics claimed, among other things, that *Voprosy filosofii* had sometimes merely described foreign ideas and not proceeded to criticize them. During the discussion, Kedrov, Arkhiptsev, and other speakers basically defended *Voprosy filosofii*'s work, and Frolov referred to his January 1974 editorial as setting the journal's tasks, giving no sign that he intended to retreat from its controversial aspects. Only one speaker, S. M. Kovalev, admonished Frolov for his January editorial, arguing that one should not try to prevent people from criticizing just because they cannot write creative philosophy articles themselves.[38]

Frolov survived the renewed pressure, but Kedrov did not. The June 1974 *Vestnik* of the Academy of Sciences reported that Kedrov had resigned as director. As usual with such *Vestnik* reporting, no timing was indicated, but his removal appears to have come sometime between February and April 1974.[39] He returned to the Institute for History of Natural Science and Technology, but this time only as head of a sector.[40] He was replaced by B. S. Ukraintsev, head of a sector at the Institute of Philosophy and vice-president of the Society of Philosophers.[41] Ukraintsev apparently had not been much involved in the infighting between conservatives and liberals, but had been linked with Kedrov's Lysenkoist foe G. V. Platonov, col-

[38] *Voprosy filosofii*, June 1974, p. 164.
[39] The March 1974 *Vestnik* (signed to press 12 March 1974), reporting Kedrov's seventieth birthday in December 1973, still listed him as Institute of Philosophy director, and he helped manage the 5 February 1974 institute discussion of *Voprosy filosofii*'s work (*Voprosy filosofii*, no. 6, 1974). However, the May 1974 *Voprosy filosofii* (p. 104) identified him with the Institute for History of Natural Science and Technology in reporting a 16–17 January 1974 meeting—but this may have reflected his position in May rather than in January.
[40] So identified in the August 1974 *Voprosy filosofii*, p. 190.
[41] So identified in the November 1972 *Voprosy filosofii*, p. 190.

laborating with him on an article in the October 1964 *Voprosy filosofii.* Ukraintsev later implicitly rejected some of Kedrov's key approaches. In a July 1977 *Voprosy filosofii* article, he declared that "exaggeration of the importance" of methods borrowed from the West "presents a serious danger for social sciences and is one of the kinds of ideological weapons in bourgeois philosophy aimed against dialectical and historical materialism." He criticized proposals that the social sciences copy the natural sciences and declared that the former could become "exact sciences" only by using their own method, based on Marxism-Leninism.[42]

Frolov successfully withstood maneuvers to oust him until March 1977. In mid-1974, *Voprosy filosofii*'s editorial board underwent an extensive shake-up. Frolov's deputy was removed and one of Trapeznikov's deputies joined the board,[43] but Frolov remained chief editor and some allies (Kedrov, Iovchuk) remained on the board as well. Frolov's position weakened somewhat in November 1974 when his former chief and patron, P. N. Demichev, was demoted to minister of culture and later removed as CC secretary for ideology. In another sign of his loss of support, the Academy of Social Sciences, which had appeared on Frolov's side in 1973, became critical in 1974, apparently in response to political pressure. The August 1975 *Voprosy filosofii* reported a 1974 academy meeting run by the conservative deputy rector, G. Ye. Glezerman (rector Iovchuk apparently was absent), which criticized *Voprosy filosofii* for failing to adhere to the "principle of *partiynost.*" Momdzhyan, who had been so outspoken in defense of the journal's campaign in 1973, made the sharpest criticism, stating that the journal had made "serious errors and had shortcomings that cause alarm among the Soviet philosophical community," and that the journal was deficient in its treatment of dialectical materialism and historical materialism. A. S. Kovalchuk "sharply" denounced *Voprosy filosofii*'s January 1974 editorial for labeling the journal's critics as "scientifically unproductive people" who hide their lack of productivity behind fake *partiynost.*[44] Pressure was intensified when a *Pravda* editorial article (19

[42] *Voprosy filosofii,* July 1977, pp. 87–88.
[43] Among the five board members dropped was M. K. Mamardashvili, who had become deputy chief editor one month after Frolov became chief editor in 1968 and had served under him for six years. Among the nine added to the board was N. V. Pilipenko, longtime head of the sector for philosophy and scientific communism of the CC's Science Section and Trapeznikov's agent in controlling philosophers.
[44] *Voprosy filosofii,* August 1975, p. 176.

179

September 1975) criticized the Institute of Philosophy and *Voprosy filosofii* and called on philosophers to fight harder against objectionable ideas. The October 1975 *Voprosy filosofii* reprinted the *Pravda* piece and in an accompanying note acknowledged that it applied to the journal. In the November 1975 *Voprosy filosofii*, N. B. Bikkenin, head of Agitprop's sector for journals, implicitly criticized Frolov's views when he attacked bourgeois foes who "drew an impassable line between ideology and propaganda on the one hand, and science and information on the other, considering ideology as some sort of absolute antithesis of the scientific approach (*nauchnost*)."

Frolov's demise finally came in early 1977, when he was removed as chief editor of *Voprosy filosofii*. The March 1977 issue listed him only as a member of the editorial board, and the April 1977 issue listed conservative V. S. Semenov as the new chief editor.[45]

Despite being ousted, both Kedrov and Frolov remain prominent and respected members of the Soviet academic community, publishing articles and delivering speeches. Kedrov, who turned seventy-five in December 1978, continues to provoke debates, although on a more purely academic level. In 1977 he launched a debate over the relationship of the abstract and concrete in dialectics. In laying out his views on this theme at a conference late in that year, he asserted that Soviet textbooks on Marxist philosophy were "largely incorrect" and poorer than texts for other sciences—assertions to which other philosophers objected.[46] He continued his argumentation on this subject in articles in the January and February 1978 *Voprosy filosofii*s, stirring a big debate among philosophy teachers.[47] Touching on more directly political subjects, he used an article on Lenin and dialectics in the July 1979 *Problemy mira i sotsializma* to attack "dogmatism in theory" and "backward thinking" (as well as liberal deviations from Marxism-Leninism), and to argue that détente was not an obstacle to revolutionary change.

As for Frolov, in return for giving up his post in *Voprosy filosofii* he was elected a corresponding member of the Academy of Sciences in December 1976, and later, in 1980, became chairman of the

---

[45] Semenov had taken over as deputy director of the Institute for Concrete Social Research during the 1971–72 purge of liberals in that institute and has appeared to be on the side of the conservatives.

[46] *Filosofskiye nauki*, no. 6, 1978, pp. 146, 150–52.

[47] For followup to the articles, see the June 1979 *Voprosy filosofii*, pp. 161–71.

B. M. Kedrov — A Soviet "Liberal"

Scientific Council for Philosophical and Social Problems of Science and Technology, a body attached to the Presidium of the academy.[48] Frolov, who turned fifty in September 1979, now writes weighty articles about world problems jointly with prestigious establishment authors.[49] While focusing on subjects that are not politically controversial, Frolov still occasionally takes a jab at Lysenko, as in his July 1978 *Voprosy filosofii* article on biology.

[48] First identified as such in the February 1980 *Voprosy filosofii*, p. 190.
[49] For example, articles with V. Zagladin, the first deputy head of the CC International Section, in the March 1978 *Problemy mira i sotsializma* ("Global problems of today and Communists"), in the 7 May 1979 *Pravda* ("Global problems of today: the search for solutions"), and *Kommunist*, no. 7, May 1979 ("Global problems and the future of humanity").

Conclusion

The story spelled out above is that of a key turn in the course of
Soviet history—Stalin's postwar decision to reject relaxation at
home and abroad and to crush incipient moderation in Soviet politics
and public life. This turn played a great role—even if mainly
negative—in defining the issues that would dominate Soviet intellectual discourse for years to come. The story as documented throws
light on a number of questions, both about this particular period and
about the nature of Soviet politics in general.

Was the defeat of moderation preordained? Clearly the development of the cold war dealt a shattering blow to moderation since the
latter depended on some relaxation of rigidity at home and of tension
internationally. The imposition of the thesis that the world was
divided into two hostile camps, not only in politics and state relations
but also in science, made the position of the moderates untenable.
However, Soviet leaders apparently did not foresee the cold war in
1946. Zhdanov, one of Stalin's closest associates, appeared to
anticipate some postwar relaxation, and various moderates at lower
levels—such as the economist Ye. S. Varga—seemed to share that
assumption. Stalin's 1946 statements indicating expectations of
peace—in addition to misleading these Soviet officials—suggest that
Stalin's attitudes changed over time also.

The reasons for adopting an increasingly hard line at home and
abroad appear to lie not only in the worsening relations with the
West but also in developing domestic economic difficulties, in Stalin's character and method of rule, and in intensifying political
competition within the Soviet leadership. The worsening food

shortages in 1946–47 and other serious weaknesses prompted Stalin to take measures to cut off the Soviet public from potentially disruptive Western influence and to stimulate nationalistic fervor and hostility to all things foreign. Stalin's inordinate fear that any of his deputies might acquire substantial authority led him to stimulate constant intrigues and vicious purges. And the competition for power among his deputies and the jockeying for positions by ambitious ideologues and scientists up and down the line influenced and intensified Stalin's decisions and helped carry his acts to extremes.

This line of argument raises the question of how closely Stalin controlled political events and what influence his deputies and their maneuverings had on the final outcome of the struggle over policies and ideas. That Stalin was the supreme authority on virtually everything is not in doubt. But his fearsome dictatorial control did not prevent those below him from engaging in serious political struggles, initiating actions, developing followings, and even taking divergent stands on issues—especially in those cases when Stalin had not made his own position clear. The picture of Soviet politics that emerges from the present study is one of constant political maneuvering over personnel and policies in the academic and cultural world, party and government apparats, and regional and central organs. Stalin's direct subordinates and their protégés were usually involved in these lower-level struggles and apparently had considerable leeway to pursue their activities and intrigues and to push their own interests. Stalin appears to have even encouraged this process, deliberately playing his deputies off against one another.

The secrecy in which Stalin's activities were cloaked makes it difficult to pin down the extent to which he supervised detailed everyday party and government decisions, but the present study can provide evidence to illuminate one key feature of Stalin's leadership style: his periodic interventions, such as his 1946 ouster of Malenkov, his 1947 condemnation of Aleksandrov's history of philosophy, his 1948 secret endorsement of Lysenko's position, his 1949 initiation of the anti-cosmopolitanism campaign. Stalin's actions usually took place behind the scenes and were only revealed obliquely by others. Key decisions were made at unannounced private meetings, such as Stalin's May 1946 meeting with Zhdanov and Kuznetsov at which he made the initial decision to demote Malenkov. Public initiatives were often left to others—for example, Zhdanov's speeches announcing the cultural crackdown or Andreyev's speech on agriculture to the February 1947 plenum. Press reports on the activities of Stalin's deputies provide much evidence to help decipher political developments for the period 1946–49, but after 1950 Stalin's direct role increased and the roles of his chief deputies become more difficult to trace, making trends murkier than ever. Whereas Zhdanov clearly led the cultural crackdown of 1946–47, Malenkov and Suslov clearly reorganized the CC apparat in 1948, and Beriya clearly fabricated the Leningrad Case in 1949, comparable moves instigated and executed by Stalin's deputies are relatively difficult to find from 1950 to 1953. Stalin himself appears to have originated the Mingrelian Case (the 1951–52 purge of Beriya's protégés in Georgia), the Crimean Case (the 1952 execution of prominent Jews) and the Doctors' Plot (the late 1952–early 1953 beginning of a large-scale anti-Semitic pogrom). In such matters, Stalin appears to have turned more than earlier to his private secretariat headed by A. N. Poskrebyshev, whose role apparently grew as that of the Politburo and Stalin's formal deputies diminished.

Did Stalin's absolute dictatorship mean that his deputies played no significant role in events and that politics involved merely the will or whim of the dictator? The evidence in the present study suggests that his subordinates influenced events and that the interplay of politics influenced the outcome, especially during the 1946–49 period. In a few cases we can even see the influences at work on Stalin to incite him to important actions: the accusations of malfeasance by Zhdanov and others which led to Malenkov's demotion in mid-1946 and the intrigues by Malenkov, Beriya, Lysenko, and others which led to the fall of Zhdanov and Voznesenskiy in 1948 and 1949. Unfortunately, the individuals or factors prompting Stalin in other cases—for example, to force the philosophy debate of 1947 or the anticosmopolitanism campaign in 1949—are still unknown and his motivations are difficult to fathom.

I believe that this book demonstrates that even during the dark days of Stalin's vicious and arbitrary personal rule, "politics" existed in the Soviet Union and that the political struggles were not solely over posts and power but also over ideas and policies. Moreover, I believe that such political struggles and policy differences can be detected and documented from a close study of the public press itself. The struggles over ideas and policies documented in this study show that even under Stalin some people in the Soviet establishment were pressing for reform and, as the Kedrov chapter seeks to demonstrate, there has been a certain continuity in the struggle between "liberals" and "moderates" on the one hand, and "conservatives" and "dogmatists" on the other, through the decades.

Members of the Secretariat and Politburo (Presidium), 1945 — 53

Chart 1a. Members, 1945 – 49

Members of the Secretariat

1945 – 46	March 1946 – 47	1948	1949
Stalin	Stalin	Stalin	Stalin
Zhdanov	Zhdanov	Zhdanov[1]	
		(– Aug 48)	
Malenkov	Malenkov[2]	Malenkov	Malenkov
	(– May 46)	(48–)	
Andreyev[3]			
(– Mar 46)			
Shcherbakov[4]			
(– 45)			
	Kuznetsov[5]	Kuznetsov	
	(Mar 46–)	(– Feb 49)	
	Popov[6]	Popov	Popov
	(Mar 46–)		(– Dec 49)
	Patolichev[7]		
	(May 46 – Mar 47)		
		Suslov[8]	Suslov
		(mid-47–)	
			Ponomarenko[9]
			(mid 48–)

Appendix 1

Members of the Politburo

Stalin	Stalin	Stalin	Stalin
Zhdanov	Zhdanov	Zhdanov[1]	
		(−Aug 48)	
Molotov	Molotov	Molotov	Molotov
Andreyev	Andreyev	Andreyev	Andreyev
Kaganovich	Kaganovich	Kaganovich	Kaganovich
Mikoyan	Mikoyan	Mikoyan	Mikoyan
Voroshilov	Voroshilov	Voroshilov	Voroshilov
Khrushchev	Khrushchev	Khrushchev	Khrushchev
Kalinin	Kalinin[10]		
	(−Jun 46)		
	Beriya[11]	Beriya	Beriya
	(Mar 46−)		
	Malenkov[12]	Malenkov	Malenkov
	(Mar 46−)		
		Voznesenskiy[13]	
		(Feb 47−Mar 49)	
		Bulganin[14]	Bulganin
		(Feb 48−)	
			Kosygin[15]
			(48−)

Candidate members of the Politburo

Shvernik	Shvernik	Shvernik	Shvernik
Beriya[11]			
(−Mar 46)			
Malenkov[12]			
(−Mar 46)			
Voznesenskiy	Voznesenskiy[13]		
	(−Feb 47)		
Shcherbakov[4]			
(−45)			
	Bulganin[16]		
	(Mar 46−Feb 48)		
	Kosygin[17]	Kosygin	
	(Mar 46−)	(−48)	

[1] A. A. Zhdanov died on 31 August 1948.

[2] G. M. Malenkov was removed as secretary in early May 1946 (see Chapter 1). His first identification back in this post was in the 21 July 1948 *Pravda*.

[3] A. A. Andreyev was dropped in the March 1946 reorganization (see Chapter 1).

[4] A. S. Shcherbakov died in May 1945, according to the *Sovetskaya istoricheskaya entsiklopediya*.

[5] A. A. Kuznetsov became secretary in March 1946 (if not before, see Chapter 1). He was removed as part of the Leningrad purge in February 1949 (this date is given in volume 51 of the second edition of the *Bolshaya sovetskaya entsiklopediya*).

[6] G. M. Popov was elected secretary at the March 1946 plenum. He was dropped in mid-December 1949.

[7] N. S. Patolichev became secretary in May 1946, according to his biography in volume 32 of the second edition of the *Bolshaya sovetskaya entsiklopediya*. He was sent to the Ukraine as agriculture secretary in March 1947.

[8] M. A. Suslov became secretary in mid-1947, and was first identified as such in the 24 September 1947 *Pravda* (see Chapter 3).

[9] P. K. Ponomarenko became secretary in mid-1948 (see Chapter 3).

[10] M. I. Kalinin died in June 1946.

[11] L. P. Beriya was raised from Politburo candidate member to full member in March 1946.

[12] G. M. Malenkov was raised from Politburo candidate member to full member in March 1946.

[13] N. A. Voznesenskiy was elected to the Politburo at the February 1947 plenum; he was dismissed from state posts in March 1949.

[14] N. A. Bulganin became a Politburo member in February 1948, according to his biography in the *Bolshaya sovetskaya entsiklopediya*, 2d edition.

[15] A. N. Kosygin was added to the Politburo in 1948, according to his biography in the *Bolshaya sovetskaya entsiklopediya*, but press sources give little indication as to exactly when (see Chapter 4).

[16] N. A. Bulganin became a Politburo candidate member in March 1946, and was promoted to full member in February 1948.

[17] A. N. Kosygin became a Politburo candidate member at the March 1946 plenum and was promoted to full member sometime in 1948.

Chart 1b. Members, 1950–53

Members of the Secretariat

1950–52	Oct. 1952–March 1953	March 1953	June 1953
Stalin	Stalin[1]		
Malenkov	Malenkov	Malenkov[2] (–14 Mar 53)	
Khrushchev[3] (Dec 49–)	Khrushchev	Khrushchev	Khrushchev
Suslov	Suslov	Suslov	Suslov
Ponomarenko	Ponomarenko[4] (–6 Mar 53)		
	Mikhaylov	Mikhaylov[5] (–14 Mar 53)	
	Aristov	Aristov[6] (–14 Mar 53)	
	Pegov[7] (–6 Mar 53)		
	Ignatov[8] (–6 Mar 53)		
	Brezhnev[9] (–6 Mar 53)		
		Pospelov[10] (6 Mar 53–)	Pospelov
		Shatalin[11] (6 Mar 53–)	Shatalin
		Ignatyev[12] (6 Mar 53–Apr 53)	

Members of the Politburo   Members of the Presidium

1950–52	Oct. 1952–March 1953	March 1953	June 1953
Stalin	Stalin[1]		
Molotov	Molotov	Molotov	Molotov
Malenkov	Malenkov	Malenkov	Malenkov
Beriya	Beriya	Beriya[13] (–Jul 53)	
Voroshilov	Voroshilov	Voroshilov	Voroshilov
Kaganovich	Kaganovich	Kaganovich	Kaganovich
Mikoyan	Mikoyan	Mikoyan	Mikoyan
Bulganin	Bulganin	Bulganin	Bulganin
Khrushchev	Khrushchev	Khrushchev	Khrushchev
Kosygin[14] (–Oct 52)			
Andreyev[15] (–Oct 52)			
	Suslov		
	Pervukhin	Pervukhin	Pervukhin
	Saburov	Saburov	Saburov
	Andrianov		
	Aristov		
	Chesnokov		

Members of the Secretariat and Politburo (Presidium)

Members of the Politburo	Members of the Presidium		
	Ignatyev		
	Korotchenko		
	Kuusinen		
	Kuznetsov		
	Malyshev		
	Melnikov		
	Mikhaylov		
	Ponomarenko		
	Shkiryatov		
	Shvernik		

	Candidate members of the Politburo (Presidium)		
Shvernik	Brezhnev	Shvernik	Shvernik
	Ignatov	Melnikov[16]	
		(−Jun 53)	
	Kabanov	Ponomarenko	Ponomarenko
	Kosygin	Bagirov[17]	
		(−Jul 53)	
	Patolichev		
	Pegov		
	Puzanov		
	Tevosyan		
	Vyshinskiy		
	Yudin		
	Zverev		

[1] I. V. Stalin died on 5 March 1953.

[2] G. M. Malenkov was released as CC secretary on 14 March 1953, to concentrate on his work as premier.

[3] N. S. Khrushchev became CC secretary in December 1949.

[4] P. K. Ponomarenko was dropped as secretary on 6 March 1953 and became minister of culture on 15 March.

[5] N. A. Mikhaylov was dropped as secretary on 14 March 1953; he was elected first secretary of Moscow on 10 March.

[6] A. B. Aristov was dropped as secretary on 14 March 1953 and was exiled to the Far East as chairman of the Khabarovsk *kray* executive committee (he was listed in the latter post for 1953–54; see his biography in volume 51 of the second edition of the *Bolshaya sovetskaya entsiklopediya*).

[7] N. M. Pegov was dropped as secretary on 6 March 1953 and named secretary of the Presidium of the Supreme Soviet.

[8] N. G. Ignatov was dropped as secretary on 6 March 1953 and elected second secretary of Leningrad *oblast* and first secretary of Leningrad city on 1 April 1953 (*Leningradskaya pravda*, 2 April 1953).

[9] L. I. Brezhnev was dropped as secretary on 6 March 1953 and named chief of the political department of the navy (*Pravda*, 7 March 1953).

[10] P. N. Pospelov was elected secretary on 6 March 1953.

[11] N. N. Shatalin was elected secretary on 6 March 1953.

[12] MGB Minister S. D. Ignatyev was elected secretary on 6 March 1953, but was removed in April after being attacked for promoting the Doctors' Plot.

[13] L. P. Beriya was removed in July 1953. The *Istoriya kommunisticheskoy partii sovetskogo soyuza*, vol. 5, bk. 2, p. 640, gives the date of the plenum that formally

removed Beriya as 2–7 July 1953.

[14] A. N. Kosygin became only a candidate member of the new Presidium elected in October 1952.

[15] A. A. Andreyev was not elected to the new Presidium formed in October 1952.

[16] L. G. Melnikov was fired as Ukrainian first secretary in June 1953 (*Pravda Ukrainy*, 13 June 1953) and appointed ambassador to Rumania in July (*Pravda*, 27 July 1953).

[17] M. D. Bagirov was ousted as Azerbaydzhan first secretary in July 1953 (*Pravda*, 19 July 1953) after the arrest of his patron Beriya.

Leading Government Posts, 1945 – 53

Chart 2a. Leading posts, 1945 – 49

Government posts	1945	1946 – 48	1949
Premier	Stalin	Stalin	Stalin
Deputy premiers	Molotov[1]	Molotov	Molotov
	Beriya[2]	Beriya	Beriya
	Kaganovich[3]	Kaganovich	Kaganovich
	(– Feb 47)	(Dec 47 –	
	Voznesenskiy[4]	Voznesenskiy	
		(– Mar 49)	
	Mikoyan[5]	Mikoyan	Mikoyan
	Kosygin[6]	Kosygin	Kosygin
		Voroshilov[7]	Voroshilov
		(Mar 46 –)	
		Andreyev[8]	Andreyev
		(Mar 46 –)	
		Malenkov[9]	Malenkov
		(Aug 46 –)	
		Saburov[10]	Saburov
		(Feb 47 –)	
		Bulganin[11]	Bulganin
		(Mar 47 –)	
		Malyshev[12]	Malyshev
		(Dec 47 –)	
			Krutikov[13]
			Yefremov[14]
			(Mar 49 –)
			Tevosyan[15]
			(Jun 49 –)
Gosplan chairman	Voznesenskiy	Voznesenskiy[4]	Saburov[16]
		(– Mar 49)	
Foreign minister	Molotov	Molotov[17]	Vyshinskiy
		(– Mar 49)	

Chart 2a. Leading posts, 1945–49

Government posts	1945	1946–48	1949
Interior minister	Beriya[18] (–Jan 46)	Kruglov	Kruglov
State security minister	Merkulov[19] (–mid-46)	Abakumov	Abakumov
Defense minister	Stalin[20] (–Mar 47)	Bulganin[21] (–Mar 49)	Vasilevskiy
Foreign trade minister	Mikoyan	Mikoyan[22] (–Mar 49)	Menshikov

[1] V. M. Molotov had been first deputy premier until the March 1946 reorganization (he was so identified, for example, in the 3 January 1946 *Moskovskiy bolshevik*), after which he was deputy premier (see *Pravda* for 20 March 1946).

[2] L. P. Beriya was already deputy premier before the March 1946 reorganization (see his identification as such in the 3 January 1946 *Izvestiya*).

[3] L. M. Kaganovich was deputy premier before the March 1946 reorganization (*Izvestiya*, 3 January 1946) but gave up this post when he was sent to the Ukraine as first secretary in February 1947. When he returned to Moscow in December 1947, he was reappointed deputy premier (by an ukase of 18 December 1947, according to the 5 February 1948 *Pravda*).

[4] N. A. Voznesenskiy was already deputy premier before March 1946 (identified in the 3 January 1946 *Izvestiya*). He was dismissed as deputy premier and Gosplan chairman in March 1949 (*Pravda*, 15 March 1949).

[5] A. I. Mikoyan was already deputy premier before March 1946 (*Izvestiya*, 3 January 1946).

[6] A. N. Kosygin was deputy premier before March 1946 (*Izvestiya*, 4 January 1946, and *Pravda*, 30 January 1946).

[7] K. Ye. Voroshilov became deputy premier in March 1946. Although Voroshilov's official biography (V. S. Akshinskiy, *Kliment Yefremovich Voroshilov*, [Moscow, 1974], p. 247) lists him as deputy premier 1940–53, he apparently lost the post during World War II. Most biographies list him as deputy premier 1946–53, and in the January 1946 nominations for the Supreme Soviet he was not identified as deputy premier even though others were.

[8] A. A. Andreyev became deputy premier in March 1946 (see Chapter 1).

[9] G. M. Malenkov became deputy premier on 2 August 1946 (*Pravda*, 19 October 1946). He was also identified as deputy premier in late 1947 (*Vechernyaya Moskva*, 4 December 1947). He apparently retained this post even after becoming CC secretary again in mid-1948, since he was identified periodically as both secretary and deputy premier during later years. But identifications of him were contradictory: he was identified as secretary and deputy premier in the 5 February 1950 *Pravda*, 19 February 1950 *Leningradskaya pravda*, 26 January 1951 *Sovetskaya Estoniya*, 8 January 1952 *Pravda*, 17 January 1953 *Moskovskaya pravda*, and 27 January 1953 *Sovetskaya Latviya*. On the other hand, he was identified only as secretary in the 30 November 1950 *Vechernyaya Moskva*, 24 January 1953 *Moskovskaya pravda*, 25 January 1953 *Sovetskaya Latviya*, and 6 February 1953 *Vechernyaya Moskva*, for example. The lack of clarity in Malenkov's status as deputy premier from 1946 to 1953 appears to have been reflected in his biography in the *Bolshaya sovetskaya entsiklopediya*, 2d ed. Mentioning his election to the Politburo in March 1946, the article vaguely states that "at this time" he was CC secretary and deputy premier, but contrary to the encyclopedia's normal practice, no dates are given.

[10] M. Z. Saburov was named deputy premier in February 1947 (see his biography in the second edition of the *Bolshaya sovetskaya entsiklopediya*).

[11] N. A. Bulganin was named deputy premier and defense minister in March 1947 (according to the 5 February 1948 *Pravda*).

[12] V. A. Malyshev was named deputy premier on 19 December 1947, according to the 5 February 1948 *Pravda*.

[13] A. D. Krutikov, only deputy minister of foreign trade in early 1948 (*Pravda*, 8 February 1948), was identified as deputy premier in the 19 December 1948 *Pravda*. He apparently did not retain this post long, for he was not even elected to the Supreme Soviet in March 1950 and hence was apparently no longer deputy premier in the new Council of Ministers formed in June 1950.

[14] A. I. Yefremov was named deputy premier on 8 March 1949 (according to the 22 March 1949 *Vedomosti verkhovnogo soveta SSSR* and 15 March 1949 *Pravda*).

[15] I. F. Tevosyan was named deputy premier on 13 June 1949 (*Vedomosti verkhovnogo soveta SSSR*, 18 June 1949).

[16] M. Z. Saburov was appointed Gosplan chairman on 5 March 1949 (*Leningradskaya pravda*, 15 March 1949).

[17] V. M. Molotov was relieved as foreign minister on 4 March 1949 and replaced with A. Ya. Vyshinskiy (*Vedomosti verkhovnogo soveta SSSR*, 22 March 1949).

[18] L. P. Beriya was replaced with S. N. Kruglov as internal affairs minister in January 1946 (*Moskovskiy bolshevik*, 15 January 1946).

[19] V. N. Merkulov was replaced with V. S. Abakumov as state security minister sometime during mid-1946 (confirmation of the change was announced at the 18 October 1946 session of the Supreme Soviet; see the 19 and 20 October 1946 *Pravda*).

[20] N. A. Bulganin succeeded Stalin as defense minister on 3 March 1947 (*Pravda*, 5 February 1948).

[21] A. M. Vasilevskiy replaced Bulganin as defense minister in March 1949 (see Vasilevskiy's biography in the second edition of the *Bolshaya sovetskaya entsiklopediya*).

[22] A. I. Mikoyan was replaced with M. A. Menshikov as foreign trade minister on 4 March 1949 (*Vedomosti verkhovnogo soveta SSSR*, 22 March 1949).

Chart 2b. Leading posts, 1950–53

Government posts	1950–51	1952	1953
Premier	Stalin	Stalin[1] (−Mar 53)	Malenkov[2]
Deputy premiers	Malenkov	Malenkov (−Mar 53)	
	Molotov	Molotov	Molotov[3]
	Beriya	Beriya	Beriya[3] (−Jul 53[4])
	Kaganovich	Kaganovich	Kaganovich[3]
	Bulganin	Bulganin	Bulganin[3]
	Mikoyan	Mikoyan	Mikoyan[5]
	Voroshilov	Voroshilov[6] (−Mar 53)	
	Andreyev	Andreyev[7] (−Mar 53)	
	Kosygin	Kosygin[8] (−Mar 53)	
	Saburov	Saburov[9] (−Mar 53)	
	Malyshev	Malyshev[10] (−Mar 53)	
	Tevosyan	Tevosyan[11] (−Mar 53)	
	Yefremov[12] (−Nov 51)		
	Pervukhin[13] (Jan 50–)	Pervukhin (−Mar 53)	
Gosplan chairman	Saburov	Saburov (−Mar 53)	Kosyachenko[14] (−Jun 53)
Foreign minister	Vyshinskiy	Vyshinskiy[15] (−Mar 53)	Molotov
Interior minister	Kruglov	Kruglov[16] (−Mar 53)	Beriya (−Jul 53)
State security minister	Abakumov[17] (−51)	Ignatyev[18] (−Mar 53)	(office abolished)
Defense minister	Vasilevskiy	Vasilevskiy[19] (−Mar 53)	Bulganin
Foreign trade minister	Menshikov	Menshikov[20] (−Mar 53)	Mikoyan

[1] I. V. Stalin died on 5 March 1953.

[2] G. M. Malenkov was appointed premier on 6 March 1953.

[3] In the reorganization after Stalin's death, four first deputy premiers were appointed on 6 March. Mikoyan was added as a deputy premier when the full list of the Council of Ministers was published on 16 March. There had been thirteen deputy premiers until March 1953 (Malenkov, Molotov, Beriya, Kaganovich, Bulganin, Mikoyan, Voroshilov, Andreyev, Kosygin, Saburov, Malyshev, Tevosyan, and Pervukhin were so identified in the 5–7 February 1953 issues of *Vechernyaya Moskva* and the 17 January 1953 *Moskovskaya Pravda*).

[4] L. P. Beriya was removed in July 1953.

[5] A. I. Mikoyan was reappointed a deputy premier on 15 March 1953.

[6] K. Ye. Voroshilov was chosen to be chairman of the Supreme Soviet Presidium on 6 March 1953.

[7] A. A. Andreyev went into semiretirement after 1953, holding only the post of member of the Presidium of the Supreme Soviet (see his biography in the third edition of the *Bolshaya sovetskaya entsiklopediya*).

[8] A. N. Kosygin lost his post as deputy premier in March 1953 but remained light industry minister.

[9] M. Z. Saburov became minister of machine building on 6 March 1953.

[10] V. A. Malyshev became minister of transport machinery and heavy machine building on 6 March 1953.

[11] I. F. Tevosyan became minister of the metallurgy industry on 15 March 1953.

[12] A. I. Yefremov died on 23 November 1951 (*Pravda*, 24 November 1951).

[13] M. G. Pervukhin was appointed deputy premier on 17 January 1950, according to the 24 January 1950 *Vedomosti verkhovnogo soveta SSSR* and the 18 January 1950 *Pravda*. He became minister of power plants and electrical equipment on 6 March 1953.

[14] M. Z. Saburov became minister of machine building on 6 March 1953, and G. P. Kosyachenko became Gosplan chairman. Kosyachenko was removed on 29 June 1953 (see Pegov's speech in the 9 August 1953 *Pravda*) and demoted to deputy chairman (identified in the 2 October 1954 *Pravda*) and replaced with Saburov.

[15] A. Ya. Vyshinskiy was demoted to first deputy minister and Molotov was appointed foreign minister on 6 March 1953.

[16] L. P. Beriya was named head of the Internal Affairs Ministry (MVD) when it and the State Security Ministry (MGB) were combined on 6 March 1953. Beriya was arrested in July 1953, however.

[17] V. S. Abakumov disappeared in late 1951 (see Chapter 5).

[18] S. D. Ignatyev became state security minister in late 1951, but was relieved in the March 1953 reorganization and became a CC secretary. He was accused of "political blindness" for permitting the development of the Doctors' Plot (6 April 1953 *Pravda*), and his removal as CC secretary was announced in the 7 April 1953 *Pravda*.

[19] N. A. Bulganin was named minister of defense and Vasilevskiy became first deputy minister on 6 March 1953.

[20] A. I. Mikoyan was appointed head of the Ministry of Domestic and Foreign Trade when the ministries of domestic and foreign trade were combined on 6 March 1953.

The Central Committee Apparat

Since the Central Committee apparat underwent several reorganizations during the period covered by this book, a brief history of the changes in the CC structure is necessary to clarify the following charts. The 1939 party congress enacted a basic change in the CC structure, rejecting the principle of having separate CC "sections" (*otdely*) to supervise each branch of the economy. Almost all sections, including these "branch" sections, were abolished or placed within one of the two new "administrations" (*upravleniya*), Cadres and Propaganda (headed by Malenkov and Zhdanov, respectively). This arrangement prevailed until 1948, although various lesser changes occurred during the war and in 1946. In 1948 Agitprop and the cadres departments were reorganized and independent industrial branch sections were reestablished. Finally, during 1952–53 the industrial sections were recombined once more and the Science Section was also reorganized.

The 1939 Reorganization

Acting upon Zhdanov's proposal to eliminate industrial branch sections, the 18th Party Congress in 1939 created an almost totally new structure based on the so-called functional principle. The party statute adopted by that congress listed the CC departments as comprising only the following: a Cadres Administration, Propaganda and Agitation Administration, Organizational-Instructor Section, Agriculture Section, and Schools Section.[1]

Despite Zhdanov's condemnation of industrial sections, such

[1] *KPSS v rezolyutsiyakh i resheniyakh syezdov, konferentsiy i plenumov TsK*, vol. 5,

sections did not really disappear, but were recreated as sections within the Cadres Administration. As the CC journal *Partiynoye stroitelstvo* explained immediately after the congress in a July editorial, the new Cadres Administration would have "separate sections for cadres of each people's commissariat," in addition to "sections for cadres of party organizations, cadres of the press and publishing houses, cadres of scientific establishments, cadres of Komsomol organizations, and cadres of trade union organizations."[2] Hence, Zhdanov's victory was not as clearcut as it had first appeared. Moreover, the reorganization centralized supervision of cadres for all industrial branches in Malenkov's hands, since Malenkov headed the Cadres Administration. Malenkov quickly emerged as a leading spokesman on industry (for example, delivering a key speech on the subject at the 18th Party Conference in February 1941) and played a central role as supervisor of much of industry during the war.

Some modifications of the early 1939 reorganization began almost immediately. Within months independent industrial sections began to be restored locally. A 29 November 1939 Politburo decree created industrial sections in republic central committees, as well as in *oblast* and city party committees,[3] and a resolution of the February 1941 18th Party Conference, acting upon Malenkov's recommendation, created republic, *oblast,* and city secretaries for branches of industry and transport.[4] Branch sections proliferated at the local level and by 1946–47, for example, the Ukrainian CC had independent sections for agriculture, the coal industry, construction, electric power stations, the petroleum industry, the light and food industry, and the metallurgical industry, judging by occasional mentions in *Pravda Ukrainy* and *Radyanska Ukraina.*

Despite the restoration of independent local industrial sections,

---

covering 1931–41, published 1971, pp. 372–73. For more on the change to the "functional" principle, see the article by L. A. Maleyko in the February 1976 *Voprosy istorii KPSS,* pp. 120–21.

[2] *Partiynoye stroitelstvo,* no. 13, July 1939, p. 6. Moreover, the editorial specified that cadres sections of republic central committees and *oblast* party committees would have the following sectors (*sektory*): for cadres of party organizations, cadres of soviet organs, cadres of industry, cadres of transport and communications, agricultural cadres, cadres of procurement organs, cadres of trade and cooperative work, cadres of NKVD and defense organizations, cadres of prosecution and judicial organs, cadres of education and culture, cadres of health, cadres of Komsomol organizations, and a sector for registration of cadres (*uchet kadrov*).

[3] *KPSS v rezolyutsiyakh i resheniyakh syezdov, konferentsiy i plenumov TsK,* vol. 5, 8th ed., 1971, p. 423.

[4] Ibid., p. 468.

however, industrial branch sections in the CC remained within Malenkov's Cadres Administration during the war. Volume five of the multivolume CPSU history, while describing in detail party reorganizations during 1939–45, fails to mention any recreation of industrial sections in the CC. Moreover, the decrees on industry and agriculture during 1938–45 published in volumes five and six of the *KPSS v rezolyutsiyakh i resheniyakh syezdov, konferentsiy i plenumov TsK* and in volume three of the *Resheniya partii i pravitelstva po khozyaystvennym voprosam* (Decisions of the party and government on economic questions) fail to mention any industrial or agricultural sections and order actions in industry and agriculture by the Cadres Administration and Organizational-Instructor Section, indicating that no independent industrial or agricultural sections then existed.[5] That industrial sections remained within the Cadres Administration is also suggested by the fact that when an independent Heavy Industry Section was created in 1948, it was staffed with some officials from the Cadres Administration who had apparently specialized in heavy industry while in the Cadres Administration.[6]

Wartime Reorganization

A number of new sections were created during the war within Agitprop and in the military and foreign affairs fields—assuming that the latter had not already existed secretly in 1939.

The evidence of changes within Agitprop is most clear. According to the multivolume CPSU history, new sections for "propaganda groups, artistic literature, cinematography, radiobroadcasting and radiofication, and art" were created in Agitprop during the war.[7] In addition, according to an article by G. P. Polozov in the September

---

[5] On the other hand, the obituary of A. V. Gritsenko in the 19 July 1978 *Selskaya zhizn* identified him as deputy head of the CC's Agriculture Section from 1939 to 1943, suggesting that an independent section existed at least until 1943.

[6] V. I. Alekseyev, a heavy industry specialist, was identified as head of a section in the Cadres Administration in a 1943 list of awards to coal industry officials (*Vedomosti verkhovnogo soveta SSSR*, 31 October 1943) and at a 1947 coal conference (*Pravda Ukrainy*, 22 June 1947). He apparently became head of the Heavy Industry Section when it was created in 1948 (see the details in the entry on Alekseyev in the charts below). V. I. Solovyev, head of a sector in the Heavy Industry Section when he died in March 1949 (*Pravda*, 20 March 1949), had also been head of a section in the Cadres Administration before the independent section was set up (he was so identified in a 21 May 1947 *Kultura i zhizn*).

[7] *Istoriya kommunisticheskoy partii sovetskogo soyuza*, vol. 5, bk. 1, p. 405.

1976 *Voprosy istorii KPSS* (pp. 119, 122), a Science Section (*Otdel nauki*) was created during the war. Later identifications indicated that it was part of Agitprop.[8] The Schools Section, listed independently in 1939, was apparently incorporated into Agitprop during the war, since postwar identifications listed it as part of Agitprop.[9] Defense industry sections apparently proliferated during the war. Historian B. A. Abramov mentions sections set up to handle "purely economic questions" in order to expedite wartime production.[10] Some war memoirs mention defense industry sections. In his memoirs about the war A. I. Shakhurin, the former people's commissar for the aviation industry, referred to "aviation sections" of the CC, suggesting that several existed just for aviation alone.[11] At some point, probably at the end of the war, these sections apparently were united into one Defense Industry Section. I. D. Serbin, later identified as head of the "Defense Industry Section,"[12] was already listed as head of an unnamed section in the 12 November 1945 *Vedomosti verkhovnogo soveta SSSR*. Tracing the defense-related sections is difficult, however, since they are treated as secret and the Defense Industry Section has only rarely been mentioned in the press in the last three or four decades. Whether they were independent of the Cadres Administration or not is also unclear.

A foreign affairs section was apparently set up during the war also, perhaps in connection with the 1943 dissolution of the Comintern. Aleksandr Chakovskiy's historical novel about the Potsdam conference mentions leaders of the International Section (*Mezhdunarodnyy otdel*) participating in a 1943 study of postwar problems.[13] A onetime Comintern employee also referred to a Foreign Department of the CC as existing in 1943–44.[14] The existence of such a section is also suggested by the biographies of B. N. Ponomarev, who

[8] For example, in the 11 May 1948 *Kultura i zhizn.*

[9] For example, in *Kultura i zhizn,* 10 August 1946.

[10] B. A. Abramov in the March 1979 *Voprosy istorii KPSS*, pp. 60–61. Abramov is head of the sector of CPSU history at the Institute of Marxism-Leninism.

[11] *Voprosy istorii,* March 1975, p. 152.

[12] The stenographic report of the 22d CPSU Congress (*XXII syezd kommunisticheskoy partii sovetskogo soyuza, stenograficheskiy otchet,* [Moscow, 1962], p. 519) identifies Serbin in this post as of 1961.

[13] *Pobeda* (see the installment published in the November 1978 *Znamya,* p. 16).

[14] Alfred Burmeister, *Dissolution and Aftermath of the Comintern: Experiences and Observations, 1937–1947,* New York: Research Program on the USSR (East European Fund, Inc.), 1955, p. 24.

became head of the present-day International Section in the 1950s. These biographies identify him as deputy head of an unnamed CC section from 1944 to 1946.[15] As a foreign affairs specialist, Ponomarev would presumably have been in a section related to foreign affairs. Whatever existed during the war was apparently augmented in December 1945. According to the 1980 CPSU history, a 29 December 1945 Politburo decision formed a "commission on foreign policy affairs" (*komissiya po vneshnepoliticheskim delam pri Politbyuro*) and gathered a group of fifty leading party officials "to train them as important political officials in the field of foreign relations."[16]

In the late stages of the war, following liberation of Latvia, Lithuania, Estonia and Moldavia, according to the 1980 CPSU history, special bureaus to lead these republics were set up in the CC in Moscow. They were comprised of CC officials and the first secretaries and premiers of the republics and their decisions were "obligatory" for the republic leaders. The bureaus, according to the history, ensured "correct combination" of national and local interests. The bureaus for Latvia, Lithuania, and Estonia were abolished in March 1947 and that for Moldavia in April 1949. The Ukrainian CC had a similar section for its Western *oblasts* from December 1945 to August 1947.[17] Suslov headed the Bureau for Lithuania from the end of 1944 until March 1946,[18] and G. V. Perov, who became secretary of the Kolkhoz Affairs Council in Moscow in October 1946, was deputy chairman, then chairman of the Bureau for Estonia from 1944 to 1946.[19]

In addition, there were ex officio sections in the CC by the end of the war. The Main Political Administration of the Red Army operated "with the rights of a military section of the CC VKP(b)" during the war, according to the multivolume CPSU history,[20] and kept this status when it was reorganized in February 1946.[21] Moreover, the

[15] *Sovetskaya istoricheskaya entsiklopediya* and the third edition of the *Bolshaya sovetskaya entsiklopediya*.

[16] *Istoriya kommunisticheskoy partii sovetskogo soyuza*, vol. 5, bk. 2, p. 63.

[17] Ibid., pp. 222–23.

[18] See Suslov's biography in the third edition of the *Bolshaya sovetskaya entsiklopediya*.

[19] See Perov's biography in vol. 51 of the 2d ed. of the *Bolshaya sovetskaya entsiklopediya*, pp. 221–22.

[20] *Istoriya kommunisticheskoy partii sovetskogo soyuza*, vol. 5, bk. 1, p. 321.

[21] The Red Army's Main Political Administration was expanded to a Main Political Administration for the Armed Forces at the end of February 1946 (in connection with

CPSU history also mentioned that the political administration of the Peoples' Commissariat of Railroads operated "with the rights of a section of the CC VKP(b)" during the war.[22]

1946 Reorganization

At the end of the war, there was considerable reorganization in the CC apparat, in part because of the transition to peace. B. A. Abramov declares that the sections set up to handle economic questions during the war were liquidated after the war because "economic functions had to be fully transferred to state and economic organs" and party organizations were to concentrate on political leadership, checking, and ideological work.[23] This reorganization may have included the amalgamation of defense-related sections into one defense industry section.

In addition, in May 1946 the Organizational-Instructor Section, part of Malenkov's cadres apparat, was replaced by a new Administration for Checking Party Organs led by Patolichev, and a group of CC inspectors was assembled.[24] Another part of this reorganization was the introduction of a new *nomenklatura* system that listed all the positions over which the CC exercised direct control. This reorganization of the *nomenklatura* is described by Abramov, who indicates that the new system was more elaborate than the old and necessitated an October 1946 conference of local cadres chiefs to explain its operation. According to Abramov, by 1 October 1947 "over 40,000 persons" were on the CC's *nomenklatura*.[25] The reorganization continued into January 1947, when the position of CC *partorg* (party organizer at local plants) was largely abolished. As Abramov explains, *partorg*s played an important role during the war but became unnecessary afterward. After 1947 they were retained only for defense plants and special projects.[26]

1948 Reorganization

The departmental structure, both in the CC and in local party

the combination of army and navy commissariats into an armed forces commissariat) and in its new form was also given "the rights of a CC section," according to Yu. P. Petrov (*Stroitelstvo politorganov, partiynykh i komsomolskikh organizatsiy armii i flota*, 1968, p. 391).

[22] *Istoriya kommunisticheskoy partii sovetskogo soyuza*, vol. 5, bk. 1, p. 280.
[23] Abramov in the March 1979 *Voprosy istorii KPSS*, pp. 60–61.
[24] See Chapter 1 for details.
[25] Abramov in the March 1979 *Voprosy istorii KPSS*, pp. 58–59.
[26] Ibid., p. 62.

committees, was thoroughly reorganized in 1948. According to the CPSU history published in 1980, the following sections were created in the CC in July 1948: Propaganda and Agitation; Party, Trade Union, and Komsomol Organs; Heavy Industry; Light Industry; Machine Building; Transport; Agriculture; Administrative and Planning-Finance-Trade; and Foreign Relations. At the same time, sections for party, trade union, and Komsomol organs, Agitprop, agriculture, and administrative and planning-finance-trade became "obligatory" for republic central committees, as well as *kray* and *oblast* party committees. The CCs of the Ukraine, Belorussia, Latvia, Lithuania, Estonia, Moldavia and some *oblast* party committees additionally had "sections for work among women."[27]

In this reorganization, the Propaganda and Agitation Administration was renamed the Propaganda and Agitation Section, and the sections within Agitprop were changed into sectors.[28] Thus, on 11 July 1948 *Kultura i zhizn* mentioned the Schools Section of the Propaganda Administration; by late 1948 it was referred to as the Schools Sector of the Propaganda Section (*Pravda,* 30 November 1948).

The Administration for Checking Party Organs and the Cadres Administration were combined into a single cadres department, the Section for Party, Trade Union and Komsomol Organs, and in the process industrial branch sections were created out of the cadres department. At least two branch sections—for heavy industry and light industry—were mentioned in the press within months of the reorganization.[29] The statements by Abramov and the 1980 CPSU history that an Agriculture Section was created in 1948 suggest that the Agriculture Section, which had been listed as an independent department after the 1939 reorganization, had been incorporated into the Cadres Administration sometime during the war.

The international section also underwent reorganization at this time. In 1948 a Foreign Relations Section (*Otdel vneshnikh snosheniy*) was created, according to Abramov. In July 1949, how-

---

[27] *Istoriya kommunisticheskoy partii sovetskogo soyuza,* vol. 5, bk. 2, p. 220. Abramov, *Voprosy istorii KPSS,* March 1979, p. 64, includes the same list of new CC sections being created in 1948, but does not specify the month. He states that the parallel reorganization of republic, *kray* and *oblast* departments occurred at the end of 1948 and at the city and *rayon* level at the start of 1949.

[28] For details on the change, see Chapter 3.

[29] *Pravda* on 19 December 1948 mentioned the Light Industry Section and on 20 March 1949 the Heavy Industry Section.

ever, it was split into two departments: A Foreign Policy Commission (*Vneshnepoliticheskaya komissiya*) and a Section for Cadres of Diplomatic and Foreign Trade Organs (*Otdel kadrov diplomaticheskikh i vneshnetorgovykh organov*).[30] The Foreign Policy Commission later—in the 1950s—became the present-day International Section.[31] Cadres abroad apparently had been handled in the Cadres Administration before the 1948 reorganization.[32] In addition to the foreign cadres section, a separate CC Commission for Travel Abroad reportedly existed in the 1950s to process applications for travel.[33] These two bodies appear to have been later combined in the Foreign Cadres Section.[34]

1950 Reorganization

A more limited reorganization occurred in December 1950, when, according to the 1980 CPSU history, Agitprop was split into four sections: Agitprop, the Science and Higher Educational Institutions

[30] Abramov in the March 1979 *Voprosy istorii KPSS*, p. 64.

[31] Émigré Michael Voslensky, who was secretary of the Academy of Sciences' disarmament commission in the 1950s and 1960s, mentions visiting an official in the CC's Foreign Policy Commission in 1952 and explains that this body later became the International Section (M. S. Voslensky, *Nomenklatura* [Vienna, 1980], p. 176). The title *Mezhdunarodnyy otdel* (International Section) was apparently first used in the press on 24 March 1960 (in *Pravda Ukrainy*). Ponomarev, identified in the 26 October 1961 *Pravda* as head of the International Section, is identified as head of an unnamed section since 1955 in his biographies in the *Bolshaya sovetskaya entsiklopediya*, 3d ed., and the 1977 *Bolshaya* yearbook, suggesting that the International Section existed in the 1950s. About 1957, a Bloc Relations Section was split off from Ponomarev's section to deal with East European regimes. The encyclopedia biographies of the first head of that section, Yu. V. Andropov, indicate that he became head of a section in 1957. The section's name was given as the Section for Ties with Communist and Workers Parties of Socialist Countries (*Otdel po svyazyam s kommunisticheskimi i rabochimi partiyami sotsialisticheskikh stran*) in the 1959 *Spravochnik partiynogo rabotnika* [Party worker's handbook], p. 500—the only time its full name has appeared in the Soviet press.

[32] According to Voslensky, a "sector for foreign cadres" had existed as part of the Cadres Administration (Voslensky, *Nomenklatura*, p. 430).

[33] Voslensky writes that a CC Commission for Travel Abroad existed at the same time as the Section for Diplomatic Cadres, apparently during the 1950s and/or 1960s, and that the latter section was headed by A. S. Panyushkin (ibid., pp. 425–26, 441). According to Panyushkin's obituary in the 14 November 1974 *Pravda*, he became head of a section in 1953. (His biography in the 1962 *Bolshaya* yearbook, however, states that he joined the CC apparat in 1954 but only became a section head in 1959.)

[34] The diplomatic cadres section has apparently only been mentioned one other time in the Soviet press—under its later name. The 17 November 1965 *Vedomosti verkhovnogo soveta SSSR* referred to the "Foreign Cadres Section" (*Otdel zagranichnykh kadrov*).

Section, the Artistic Literature and Art Section, and the Schools Section.[35] The Schools Section first appeared in the press in the 28 February 1951 *Kultura i zhizn,* and the Section for Science and Higher Educational Institutions first was mentioned in *Sovetskaya Belorussiya* on 16 October 1951.

1952–53 Reorganization

Sometime during late 1952 or early 1953, there was a major reorganization of CC sections and of parallel sections in republic CCs, primarily affecting the industrial branch sections and the Science Section. The only evidence of the reorganization is the change in titles of these departments upon their occasional mentions in the press. The reorganization in Moscow appears to have occurred before Stalin's death or in the days immediately following, whereas the change in the republics appears to have occurred in spring or summer of 1953.

The CPSU CC structure that existed in late 1952 and early 1953 is somewhat unclear because few identifications appeared in the press. The 19th Congress did not reveal what sections existed at that time; however, some idea can be gained from observing the structure of the CC apparat in republic CCs. Several republics, in announcing the election of new party leadership bodies at the end of their September 1952 party congresses, also listed the sections of their CCs. These republics—Kazakhstan, Belorussia, Latvia, Turkmenia, Tadzhikistan, and Kirgizia—used identical names for their sections:[36] Administrative; Agitprop; Agriculture; Artistic Literature and Art; Heavy Industry (except for Kirgizia); Light Industry (except for Tadzhikistan, where the section was named the Industry Section); Party, Trade Union, and Komsomol Organs; Planning-Financial-Trade; Schools; Science and Higher Educational Institutions; Transport; and Work among Women. Scattered identifications in the Ukrainian press during 1950–53 indicated the existence of similar sections in that republic: Agitprop; Agriculture; Artistic Literature and Art;

---

[35] *Istoriya kommunisticheskoy partii sovetskogo soyuza,* vol. 5, bk. 2, p. 221.

[36] See *Kazakhstanskaya pravda, Sovetskaya Belorussiya,* and *Sovetskaya Kirgiziya* for 25 September 1952, and *Sovetskaya Latviya, Turkmenskaya iskra,* and *Kommunist Tadzhikistana* for 24 September 1952. Some had additional sections as well: Belorussia had a Machine Building Section; Latvia a Fish Industry Section; Turkmenia, Belorussia, and Kazakhstan had construction sections. Sections were not listed at the Lithuanian, Estonian, Moldavian, Armenian, Azerbaydzhani, Uzbek, Ukrainian, and Georgian congresses.

Heavy Industry; Light Industry; Party, Trade Union, and Komsomol Organs; Planning-Financial-Trade; Schools; Science and Higher Educational Institutions; Transport; Work among Women; plus Administrative Organs, City Economy, Construction, and Machine Building.

In late 1952 or early 1953 the CPSU sections for heavy industry, light industry, machine building, and transport appear to have been combined into an Industry-Transport Section, and the Science and Higher Educational Institutions Section changed into a Science and Culture Section. The biography of I. I. Kuzmin in volume 51 of the second edition of the Bolshaya sovetskaya entsiklopediya states that in 1952 he became deputy head, then head of the Industry-Transport Section and later head of the Machine Building Section.[37] The Kuzmin biography is reinforced by A. P. Panin's obituary in the 27 June 1979 Sotsialisticheskaya industriya, which listed Panin as head of a sector, then deputy head of the Machine Building Section, then deputy head of the Industry-Transport Section between 1948 and 1954. Biographies of A. M. Rumyantsev indicate that he became head of the Science Section in 1952. His biography in the third edition of the Bolshaya sovetskaya entsiklopediya lists him as head of the Science Section starting in 1952, whereas his biography in the economic journal Eko (no. 2, 1974, p. 42) listed him as head of the Science and Culture Section starting in 1952. The first press mention of the Science and Culture Section came in the 21 April 1953 Pravda, indicating that the change occurred either under Stalin or immediately upon his death. Encyclopedia dates are occasionally imprecise or inaccurate, but it appears clear that Rumyantsev had moved to Moscow by late 1952 (witness his election to CC membership and other signs of new prominence and his articles in the Moscow press). The reorganization could, of course, have occurred after Rumyantsev already had become head of the section.

A parallel reorganization surfaced in the republic press in 1953 and 1954. Whereas in 1952 and early 1953 all references to the sections in question were to separate industrial branch sections, a Science and Higher Educational Institutions Section, and a separate Administrative Section and Planning-Finance-Trade Section, by late

[37] By 1954–55 the Industry-Transport Section had again been broken up into sections for construction, heavy industry, machine building, and so on. For example, A. P. Rudakov is identified as head of the Heavy Industry Section starting 1954 in his biography in the Ukrainska radyanska entsiklopediya.

1953 or early 1954 all references to these republic sections used the names Industry-Transport, Science and Culture, and Administrative-Trade-Financial. One republic—Uzbekistan—gave some hint of the timing of the reorganization. In reporting on a 29 May 1953 Uzbek CC plenum, *Pravda vostoka* explained that a number of Uzbek CC sections were being combined and that for this reason new heads were being appointed for the Science and Culture Section, Industry and Transport Section, and Administrative-Trade-Financial Section (*Pravda vostoka,* 30 May 1953). In moves probably reflecting a similar reorganization, Lithuania and Latvia named new heads for several sections in June, including the Science and Culture, Industry-Transport, and Administrative-Trade-Financial sections (*Sovetskaya Litva,* 18 June 1953, and *Sovetskaya Latviya,* 28 June 1953). Whereas this appears to have been Latvia's first press mention of a Science and Culture Section, a Lithuanian Science and Culture Section had been referred to already in a 28 May 1953 *Sovetskaya Litva.* Latvia had used some of the old names for sections as late as 1 April 1953, when *Sovetskaya Latviya* referred to the Planning-Financial-Trade Section and Administrative Section. New sections also showed up at some of the republic party congresses in early 1954. Belorussia, which still had a Science and Higher Educational Institutions Section on 7 February 1953 (*Sovetskaya Belorussiya*), had a Science and Culture Section by early 1954, as well as an Industry-Transport Section and Administrative-Trade-Financial Section (*Sovetskaya Belorussiya,* 14 February 1954). The Turkmen congress listed Science and Culture and Administrative-Trade-Financial sections—but no industry section at all (*Turkmenskaya iskra,* 17 February 1954). Most republics did not identify their sections at either the 1952 or 1954 republic congresses, and in these cases comparisons are not possible. Among those not reporting was the Ukraine; however, it also eventually changed its names. *Pravda* on 4 March 1953 referred to a Ukrainian Science and Higher Educational Institutions Section and *Radyanska Ukraina* on 14 May 1953 referred to S. V. Chervonenko as head of the Section for Social Sciences and Higher Educational Institutions of the Ukrainian CC, but by 20 January 1955 he was being referred to in *Pravda Ukrainy* as head of the Science and Culture Section.

The following charts depict the CC administrations and sections mentioned in the press as having existed during the postwar period, with the officials identified as heading them and the appropriate press citations.

Chart 3a. Changes in CC structure

1945	1946	1948	1949–51	1952–53
Agitprop	Agitprop	Agitprop	Agitprop	Agitprop
			Schools	Schools
			Science, Education	Science, Culture
			Literature	Literature
Organizational-Instructor	Checking	Party, Trade Union, Komsomol Organs	Party, Trade Union, Komsomol Organs	Party, Trade Union, Komsomol Organs
Cadres	Cadres			
International	Foreign Affairs Commission	Foreign Relations	Foreign Policy Commission	Foreign Policy Commission
			Foreign Cadres	Foreign Cadres
Defense Industry	Defense Industry	Defense Industry	Defense Industry	Defense Industry
		Agriculture	Agriculture	Agriculture
(Economic sections apparently within Cadres Administration)		Heavy Industry	Heavy Industry	
		Light Industry	Light Industry	Industry, Transport
		Machine Building	Machine Building	
		Transport	Transport	
		Planning, Finance, Trade	Planning, Finance, Trade	Planning, Finance, Trade

Chart 3b. Agitprop leadership, 1946–48

Posts	1946	1947	1948
Agitprop chief	Aleksandrov[1]	Aleksandrov (−mid-47)	Suslov[2] (late 47−)
Deputy chiefs	Fedoseyev[3]	Fedoseyev (−mid-47)	Shepilov[4] (late 47−)
	Iovchuk[5] (−Mar 47)		Ilichev[6] (Jan 48−)
	Yegolin[7]	Yegolin (−48)	
	Kuzakov[8]		
	Krapivin[9] (−46)		
		Grigoryan[10] (46−)	

Agitprop sections		Heads of sections	
Propaganda	Kovalev[11]	Kovalev (−48)	
Agitation	Kalashnikov[12]	Kalashnikov	Kalashnikov
Central Papers[13]			
Oblast, kray, and Republic Papers	Shumilov[14]		
Local Papers		Kuroyedov[15]	Kuroyedov
Publishing Houses			Morozov[16]
Schools	Yakovlev[17]	Yakovlev	Dubrovina[18]
Science			Zhdanov[19]
Art	Lebedev[20]	Lebedev	
Cultural-Educational Establishments		Fomichev[21]	Fomichev
Unnamed sections	Stepanov[22]	Stepanov (−48)	
	Gorodetskiy[23]		
	Oleshchuk[24]		
	Satyukov[25]		
		Kuzmin[26]	

[1] G. F. Aleksandrov was identified as chief of the Propaganda and Agitation Administration in the 8 January 1946 *Pravda Ukrainy*. His appointment to this post had been announced in the 7 September 1940 *Pravda*. He was last identified as such in the 30 June 1947 *Kultura i zhizn*. He became director of the Institute of Philosophy in 1947, according to his biography in the *Filosofskaya entsiklopediya* (1960).

[2] M. A. Suslov was first identified as chief of Agitprop in the 23 November 1947 *Moskovskiy bolshevik*.

[3] P. M. Fedoseyev was identified as deputy chief of Agitprop in the 13 January 1946 *Pravda Ukrainy*. He was last identified in this post in the 13 January 1947 *Pravda*, although he was also referred to as an Agitprop official during the June 1947 philosophy debate. The *Bolshaya sovetskaya entsiklopediya* yearbook for 1971 lists him in the CC apparat from 1941 to 1947.

[4] D. T. Shepilov was first identified as deputy chief of Agitprop in the 21 November 1947 *Izvestiya* and was last identified in this post in the 21 May 1948 *Pravda*. After

the 1948 reorganization, he became head of the Propaganda and Agitation Section.

⁵ M. T. Iovchuk was identified as deputy chief of Agitprop in the 18 June 1946 *Pravda*, as well as earlier, in a 9 August 1944 CC decree on Tataria's ideological work printed in *KPSS v rezolyutsiyakh i resheniyakh syezdov, konferentsiy i plenumov TsK*, vol. 6, 1941–54, published 1971, p. 113. His biography in the third edition of the *Bolshaya sovetskaya entsiklopediya* indicates that he had become deputy chief of Agitprop in 1944. He was last identified in this post in the 31 January 1947 *Kultura i zhizn* and in *Partiynaya zhizn*, no. 3, February 1947, before being transferred to secretary of the Belorussian CC on 7 March 1947.

⁶ L. F. Ilichev was first identified as deputy chief of Agitprop in the 24 March 1948 *Pravda*. On 4 December 1947, *Vechernyaya Moskva* still identified him as chief editor of *Izvestiya*. Ilichev was apparently transferred to Agitprop in January 1948, since the obituary of his successor as *Izvestiya* chief editor, K. A. Gubin, stated that Gubin had become *Izvestiya* editor in January 1948 (*Izvestiya*, 29 April 1979). Ilichev was last identified as deputy chief of the Propaganda and Agitation Administration in the 21 May 1948 *Kultura i zhizn*. Thereafter, he was deputy head of the Propaganda and Agitation Section. The 1962 yearbook of the *Bolshaya sovetskaya entsiklopediya* lists him in the CC apparat from 1948 to 1949, then as first deputy editor of *Pravda*.

⁷ A. M. Yegolin was identified as deputy chief of Agitprop in the 20 August 1946 *Kultura i zhizn* and also as late as 5 March 1947 in *Pravda*. The 25 September 1948 *Literaturnaya gazeta* identified him as director of the Institute of Literature.

⁸ K. S. Kuzakov was identified as deputy chief of Agitprop in an October 1946 decree on OGIZ, published in the 1971 *KPSS v rezolyutsiyakh i resheniyakh syezdov, konferentsiy i plenumov TsK*. He had been included in a list of CC officials in the 12 November 1945 *Vedomosti verkhovnogo soveta SSSR*.

⁹ A. N. Krapivin's obituary in the 15 January 1947 *Pravda* identified him as deputy chief of Agitprop starting in 1945, but indicated that at some unspecified date between then and his death in 1947 he had transferred to become assistant to a CC secretary. He probably had left Agitprop before mid-1946, therefore.

¹⁰ V. G. Grigoryan, who left the post of *Zarya vostoka* chief editor in April 1946 (last listed as editor in *Zarya vostoka*, 27 April 1946), began writing articles for Agitprop's organ *Kultura i zhizn* in mid-1946 and was identified as deputy chief of Agitprop in the 11 January 1947, 11 February 1947, and 30 April 1947 issues of *Kultura i zhizn*, as well as the 8 March 1947 *Pravda*. He continued in the apparat in unspecified positions, for example, being listed as head of an unnamed CC section in 1950 (*Vechernyaya Moskva*, 30 November 1950).

¹¹ S. M. Kovalev was identified as head of Agitprop's Propaganda Section in the 30 August 1946 *Kultura i zhizn*. He was last identified in this post in the 21 April 1948 *Kultura i zhizn*. Historian Aleksandr Nekrich mentions that S. M. Kovalev—the same Kovalev who wrote the 1968 *Pravda* article justifying the invasion of Czechoslovakia—was acting director of *Gospolitizdat* in 1948 or 1949 (Nekrich, *Otreshis ot strakha*, p. 97). A Kovalev with no initials given was identified as director of *Goskultprosvetizdat* (the State Cultural and Educational Publishing House) in the 30 May 1948 *Kultura i zhizn*.

¹² K. F. Kalashnikov was identified as head of Agitprop's Agitation Section in the 10 December 1946 *Kultura i zhizn* and as head of the Mass Agitation Section as late as 11 January 1948 in *Pravda Ukrainy*.

¹³ An Agitprop Section for Central Newspapers was mentioned in the 31 March 1948 *Kultura i zhizn*.

¹⁴ N. Shumilov was identified as head of Agitprop's Section for *Oblast, Kray*, and Republic Papers in the 20 December 1946 *Kultura i zhizn*, and as head of an unnamed Agitprop section in *Partiynaya zhizn*, no. 6, March 1947.

¹⁵ V. A. Kuroyedov was identified as head of an unnamed section of Agitprop in

Partiynaya zhizn, no. 3, February 1947, as head of the Section for *Rayon* and City Papers in the 2 March 1947 *Pravda*, and as head of the Section for Local Newspapers in the 11 April 1948 *Kultura i zhizn*, and for the last time in the 10 October 1948 *Kultura i zhizn*. By 30 November 1949 *Kultura i zhizn* was identifying him as secretary of Sverdlovsk *oblast*.

[16] M. A. Morozov was identified as head of Agitprop's Publishing Houses Section on 21 May 1948 in *Pravda* and *Kultura i zhizn*, and also on 30 June 1948 in *Kultura i zhizn*.

[17] N. N. Yakovlev was identified as head of Agitprop's Schools Section in the 24 April 1946 *Moskovskiy bolshevik* and also in the 23 May 1946 and 5 July 1946 *Pravda*. He was identified in this post as late as 3 July 1947 in *Pravda*. The 21 June 1949 *Kultura i zhizn* listed him as director of the *Izdatelstvo sovetskogo khudozhnika* (Soviet Artist Publishing House).

[18] L. V. Dubrovina, identified as director of *Detgiz* (the Children's Literature Publishing House) in the 6 May 1947 *Pravda*, was identified as head of Agitprop's Schools Section in the 11 July 1948 *Kultura i zhizn*.

[19] Yu. A. Zhdanov was identified as head of Agitprop's Science Section in the 22 April 1948 *Pravda*.

[20] P. I. Lebedev was identified as head of Agitprop's Art Section in the 7 November 1946 *Kultura i zhizn*. He had left by late 1948, since the 3 August 1948 *Pravda* identified him as chairman of the Art Affairs Committee. Poor leadership was cited when he was fired as chairman on 24 April 1951 (*Vedomosti verkhovnogo soveta SSSR*, 12 May 1951).

[21] V. Fomichev was identified as head of an unnamed Agitprop section in the 21 May 1947 *Kultura i zhizn* and as head of Agitprop's Section for Cultural-Educational Establishments as late as 26 March 1948 in *Pravda*.

[22] V. P. Stepanov was identified as head of an unnamed Agitprop section in the 20 October 1946 *Kultura i zhizn*. His biography in the 1962 yearbook of the *Bolshaya sovetskaya entsiklopediya* listed him as chief editor of *Gospolitizdat* starting in 1946, as a member of the CC apparat for 1945–48, and as deputy editor of *Kultura i zhizn* starting in 1949.

[23] Ye. N. Gorodetskiy was identified as head of an unnamed Agitprop section in the 30 August 1946 *Kultura i zhizn*.

[24] F. N. Oleshchuk was identified as head of an unnamed Agitprop section in the 20 October 1946 *Kultura i zhizn*.

[25] P. A. Satyukov was listed as head of a section before becoming *Kultura i zhizn* deputy editor (in 1946)—see his biography in the 1962 *Deputaty verkhovnogo soveta SSSR*.

[26] L. F. Kuzmin was identified as head of an unnamed Agitprop section (*Pravda*, 29 January 1947, and 7 February 1947).

Chart 3c. Leadership in other departments of the CC, 1946–48

	1946	1947	1948
Cadres Aministration			
Chief	Malenkov[1] (–mid-46)		
Deputy chief	Shatalin[2]		
	Nikitin[3]	Nikitin	
	Pavlenko[4]	Andreyev	Andreyev
	Andreyev[5]	Larionov[6] (Sep 46–)	Larionov (–Dec 48)
		Mironov[7]	
Checking Administration (created May 1946)			
Chief	Patolichev[8] (May 46–Mar 47)		
First deputy chief	Andrianov[9]	Andrianov	
Deputy chief	Ignatyev[10] (–Mar 47)		
	Borkov[11] (mid-46–)	Borkov	Borkov (–48)
		Pegov[12]	Pegov (–48)
Organizational-Instructor Section[13]			
Head	Shamberg[14] (–Mar 46)	(office abolished)	(office abolished)
	Patolichev[15] (–May 46)		
Administrator of Affairs	Krupin[16]	Krupin	Krupin
Defense Industry Section	Serbin[17]	Serbin	Serbin
Foreign Policy Affairs Commission			
Special Sector	Poskrebyshev[18]	Poskrebyshev	Poskrebyshev

[1] In March 1939 G. M. Malenkov was named chief of the Cadres Administration, which he apparently continued to head until his demotion in 1946, although there were no identifications during 1946.

[2] N. N. Shatalin was identified as deputy chief of the Cadres Administration in the 4 January 1946 *Izvestiya* and the 12 January 1946 *Pravda*. *Vechernyaya Moskva* listed him on 4 December 1947 as chief editor of *Partiynaya zhizn* and as a CC inspector. His last identification as deputy chief of the administration was in the 6 August 1946 *Pravda Ukrainy*.

[3] V. D. Nikitin was identified as deputy chief of the Cadres Administration in the 12 January 1946 *Izvestiya*. According to his 18 April 1959 *Pravda* obituary, he was deputy chief of this administration starting in 1944. He was last identified as deputy chief of the Cadres Administration in *Vechernyaya Moskva* on 4 December 1947.

[4] A. S. Pavlenko was identified as deputy chief of the Cadres Administration in the 2 November 1946 *Pravda*, 7 November 1946 *Kultura i zhizn,* and 23 December 1946 *Pravda.*

[5] Ye. Ye. Andreyev, identified only as a CC official in the lists of officials given in the 7 November 1945 *Pravda* and 12 November 1945 *Vedomosti verkhovnogo soveta*

SSSR, was specified as deputy chief of the Cadres Administration in *Partiynaya zhizn*, no. 2, November 1946. He was last identified in this post in the 4 December 1947 *Vechernyaya Moskva*. He was identified as party secretary of Gosplan in the 5 February 1953 *Vechernyaya Moskva*.

[6] A. N. Larionov was released as Yaroslavl first secretary in August 1946 to transfer to Moscow (*Pravda*, 28 August 1946) and was first identified as deputy chief of the Cadres Administration in the 2 November 1946 *Pravda* and 7 November 1946 *Kultura i zhizn*. His 24 September 1960 *Pravda* obituary lists him as deputy chief of the Cadres Administration starting in September 1946. The obituary also indicates that he was transferred to first secretary of Ryazan in December 1948.

[7] N. M. Mironov was identified as deputy chief of the Cadres Administration in the 12 January 1947 *Pravda* and the 4 December 1947 *Vechernyaya Moskva*.

[8] N. S. Patolichev was named chief of the new Checking Administration on 4 May 1946, according to his autobiography. He was transferred to the Ukraine as secretary in March 1947.

[9] V. M. Andrianov was named first deputy chief of the new Checking Administration in mid-1946, according to Patolichev's autobiography. He was never identified in this post in the press. He became Leningrad first secretary in February 1949.

[10] S. D. Ignatyev was first identified as deputy chief of an unnamed administration in the 9 October 1946 *Pravda* and was specified as deputy chief of the Checking Administration in *Partiynaya zhizn*, no. 3. February 1947, and the 1 February 1947 *Pravda*. He left to become Belorussian CC secretary on 7 March 1947.

[11] G. A. Borkov's release as Kazakhstan first secretary was announced in the 23 June 1946 *Kazakhstanskaya pravda*, and his first identification as deputy chief of the Checking Administration appeared in the 22 January 1947 *Pravda*. His last identification as deputy chief of the Checking Administration was in the 11 June 1948 *Pravda*. On 11 September 1949, *Kultura i zhizn* referred to him as Saratov *oblast* secretary.

[12] N. M. Pegov was first identified as deputy chief of the Checking Administration in the 21 May 1947 *Kultura i zhizn* and in *Partiynaya zhizn*, no. 10, May 1947. Last identified as deputy chief of the Checking Administration in the 11 June 1948 *Pravda*, he was identified as head of the new Light Industry Section in the 19 December 1948 *Pravda*.

[13] The Organizational-Instructor Section was replaced by the Checking Administration in May 1946.

[14] M. A. Shamberg was identified as head of the Organizational-Instructor Section in the 15 December 1945 *Pravda*. He was succeeded by Patolichev at the time of the March 1946 plenum.

[15] N. S. Patolichev was appointed head of the Organizational-Instructor Section at the time of the March 1946 plenum, according to Patolichev's autobiography.

[16] D. V. Krupin, listed as a CC official in the 12 November 1945 *Vedomosti verkhovnogo soveta SSSR*, was identified as Administrator of Affairs in the 8 January 1947 *Pravda* and 4 December 1947 *Vechernyaya Moskva*.

[17] I. D. Serbin was identified as head of an unnamed section in the 12 November 1945 *Vedomosti verkhovnogo soveta SSSR*, as well as in the 30 November 1950 *Vechernyaya Moskva*, although his later biographies list him as a section head starting only in 1958. Only later—in the stenographic record of the 22nd CPSU Congress in 1961—was the section headed by him identified as the Defense Industry Section.

[18] A. N. Poskrebyshev was listed as head of the special sector when nominated for the Moscow city soviet in December 1947 (*Vechernyaya Moskva*, 4 December 1947).

Chart 3d. Leadership of CC apparat, 1948–49

	1948	1949
Agitprop Section[1]		
Head	Shepilov[2]	Shepilov
	(48–)	(–49)
Deputy heads	Ilichev[3]	
	(–49)	
	Kalashnikov[4]	Kalashnikov
	(48–)	
	Slepov[5]	Slepov
	(48–)	
	Golovenchenko[6]	Golovenchenko
	(48–)	
		Kuznetsov[7]
		(–50)
		Popov[8]
		(49–)
		Kruzhkov[9]
		(49–)
Head of Agitprop sector for		
party propaganda	Lyashchenko[10]	Lyashchenko
local papers	Kuroyedov[11]	
	(–49)	
schools	Dubrovina[12]	
	(–49)	
science	Zhdanov	Zhdanov
Party, Trade Union, and Komsomol Organs Section[13]		
Head		
Deputy heads	Dedov[14]	Dedov
	(48–)	
Administrator of Affairs[15]	Krupin	Krupin
Agriculture Section[16]	Kozlov	Kozlov
Defense Industry Section	Serbin	Serbin
Foreign Relations Section[17]		(abolished)
Foreign Policy Commission		
Foreign Cadres Section		
Heavy Industry Section[18]		
Light Industry Section	Pegov[19]	Pegov
Machine Building Section		
Planning-Finance-Trade Section		
Transport Section		
Special Sector	Poskrebyshev	Poskrebyshev

[1]The Propaganda and Agitation Administration became the Propaganda and Agitation Section in mid-July 1948. The administration was last mentioned in the 11 July 1948 *Kultura i zhizn*, whereas the section was first mentioned in the 21 July 1948 issue of the same paper.

[2] D. T. Shepilov, the deputy chief of the old administration, was first identified as head of the new section in the 31 July 1948 *Pravda*. He was last identified as head in

the 30 March 1949 *Literaturnaya gazeta,* although the 13 July 1949 CC decree also referred to him as such (*Pravda,* 24 December 1952). He was shortly afterward removed, however, and *Pravda* on 23 February 1950 identified him only as a CC inspector.

[3] L. F. Ilichev became a deputy head of the new Agitprop section and was last identified as such in the 6 May 1949 *Pravda.* According to the 1962 yearbook of the *Bolshaya sovetskaya entsiklopediya,* he became first deputy *Pravda* editor in 1949.

[4] K. F. Kalashnikov was identified as deputy head of Agitprop in the 3 August 1948 *Pravda,* 11 August 1948 *Kultura i zhizn,* and 3 January 1949 *Pravda.* Before the 1948 reorganization, he had been head of the Agitation Section in Agitprop.

[5] L. A. Slepov was identified as deputy head of Agitprop in the 31 July 1948 *Pravda* and *Kultura i zhizn.* Before the reorganization, he had been editor of *Pravda*'s party life department (*Kultura i zhizn,* 11 April 1948). His obituary in the 27 October 1978 *Moskovskaya pravda* indicates that he was deputy head of Agitprop from 1948 to 1952.

[6] F. Golovenchenko was identified as deputy head of Agitprop in the 31 December 1948 *Kultura i zhizn.* He had been listed as director of *Goslitizdat* (the State Publishing House for Artistic Literature) in the 11 March 1947 *Kultura i zhizn.* According to émigré Michael Voslensky, Golovenchenko was named deputy head of Agitprop specifically to lead the anticosmopolitanism campaign (Voslensky, *Nomenklatura,* p. 415). Voslensky writes that he had been a professor of literature at the university where Voslensky wrote his dissertation. He describes a 1949 speech by Golovenchenko attacking Jews for trying to take key jobs.

Golovenchenko apparently overreached himself by attacking writer Ilya Ehrenburg, a Jew still in Stalin's favor. Ehrenburg, in his *Post-War Years 1945–1954* (pp. 132–33), relates that starting February 1949, apparently in connection with the campaign against cosmopolitanism, he (Ehrenburg) was not permitted to publish anything, and his status was placed in doubt because he was Jewish. At the end of March a friend told him that the day before, at a lecture on literature, "a speaker who at the time held a rather responsible position had announced in the presence of over a thousand people: 'I can give you some good news: Cosmopolitan Number One and enemy of the people Ilya Ehrenburg has been exposed and arrested.'" Ehrenburg, who had in fact not been arrested, wrote to Stalin asking about this statement, after which Malenkov called to reassure him. Soon editors were welcoming Ehrenburg's writings again.

The speaker who attacked Ehrenburg was apparently Golovenchenko. In a 7 July 1976 Radio Liberty Russian-language research report ("Literatura nravstvennogo soprotivleniya," pt. 4, "Stalinskiy prigovor Vasiliyu Grossmanu," RS 341/76), émigré Soviet writer Grigoriy Svirskiy tells the same story (although he dates it in 1952). Svirskiy explains that Golovenchenko, whom he identifies as head of the CC's Culture Section, announced that "enemy of the people, cosmopolitan no. one" Ehrenburg had been arrested. Svirskiy claims that, after finding out that Ehrenburg was still free and in Stalin's favor, Golovenchenko suffered a heart attack.

[7] A. N. Kuznetsov was identified as deputy head of Agitprop in the 19 January 1949 *Pravda.* His obituary in the 2 March 1974 *Izvestiya* listed him as graduating from the Academy of Social Sciences in 1950 and becoming head of Lithuania's Agitprop at that time.

[8] D. M. Popov was identified as deputy head of Agitprop in the 31 January 1951 *Kultura i zhizn.* His obituary in the 9 January 1952 *Pravda* lists him as Smolensk first secretary 1940–48, then deputy head of Agitprop starting in 1949.

[9] V. S. Kruzhkov was identified as deputy head of Agitprop in the 24 February 1950 *Pravda.* He had been identified as director of the Marx-Engels-Lenin Institute in the 15 January 1949 *Pravda,* but was replaced by Pospelov during 1949.

[10] P. N. Lyashchenko was identified as head of the party propaganda sector of Agitprop in the 31 July 1949 *Kultura i zhizn.*

[11] V. A. Kuroyedov, who headed the corresponding section in the old administration, was identified as head of the local newspapers sector of Agitprop in the 19 August 1948 *Pravda* and 10 October 1948 *Kultura i zhizn.* The 30 November 1949 *Kultura i zhizn,* however, listed him as Sverdlovsk *oblast* secretary.

[12] L. V. Dubrovina, who had headed the Schools Section in the old administration, was identified as head of the schools sector in the new Agitprop in the *Pravda*s of 20 August and 30 November 1948. But on 31 January 1950 *Kultura i zhizn* listed her as deputy minister of education of the RSFSR.

[13] The Cadres Administration and Administration for Checking Party Organs were merged into the Party, Trade Union, and Komsomol Organs Section in late 1948.

[14] A. L. Dedov was identified as deputy head of the Section for Party, Trade Union, and Komsomol Organs in the 21 February 1950 *Pravda.* His biography in the 5 March 1950 *Sovetskaya Belorussiya* lists him as deputy head of this section since 1948.

[15] An Administrative Section was created in 1948, according to B. A. Abramov's statement in the March 1979 *Voprosy istorii KPSS* (p. 64). It presumably was headed by Krupin, who was identified as Administrator of Affairs (*Upravlyayushchiy delami*) of the CC both before and after 1948.

[16] An Agriculture Section was created during the 1948 reorganization, according to Abramov (*Voprosy istorii KPSS,* March 1979, p. 64), although some sort of agriculture section surely existed before this. An independent Agriculture Section had been listed at the 18th Congress, but, judging by Abramov's statement and the apparent inclusion of branch sections in the Cadres Administration, it was probably placed in the Cadres Administration, along with branch industrial sections, in the early 1940s. A. I. Kozlov probably headed it already before the 1948 reorganization, since he was referred to by Khrushchev as head of the Agriculture Section during the 1947–48 period (see Khrushchev's 2 November 1961 speech, in Khrushchev's speeches, vol. 6, p. 58). Moreover, Kozlov's career suggests appointment to this post in early 1947. An A. Kozlov had been referred to in a January 1946 *Partiynoye stroitelstvo* (no. 1) as head of a sector of the Organizational-Instructor Section. Next, A. I. Kozlov was named minister of livestock raising in March 1946 (*Pravda,* 27 March 1946). When the ministry was combined with the Ministry of Farming on 4 February 1947 to form the Ministry of Agriculture (*Pravda,* 5 February 1947), Kozlov dropped out of sight and apparently at that time was transferred to head the CC Agriculture Section.

[17] A Foreign Relations Section (*Otdel vneshnikh snosheniy*) was created in 1948, according to B. A. Abramov in the March 1979 *Voprosy istorii KPSS* (p. 64). In July 1949, according to the same source, it was replaced by a Foreign Policy Commission (*Vneshnepoliticheskaya komissiya*) and a Section for Cadres of Diplomatic and Foreign Trade Organs (*Otdel kadrov diplomaticheskikh i vneshnetorgovykh organov*).

[18] On 20 March 1949, *Pravda* mentioned the death of an official of the Heavy Industry Section (V. I. Solovyev, head of a sector of this section). According to Abramov in the March 1979 *Voprosy istorii KPSS,* this section, as well as branch sections for light industry, machine building, and transport, was set up in 1948. Before the reorganization, Solovyev had been head of a section in the Cadres Administration (so identified in the 21 May 1947 *Kultura i zhizn*).

[19] N. M. Pegov was identified as head of the Light Industry Section in the 19 December 1948 *Pravda.*

Appendix 3

Sections of CC	1950–51	1952–53	1953–54
Agitprop			
Head		Mikhaylov[1]	Kruzhkov[2]
Deputy heads	Kruzhkov[2]	Kruzhkov (–53)	
	Stepanov[3] (51–)		
	Slepov[4] (–52)		
	Popov[5] (–Jan 52)		
Party, Trade Union, and Komsomol Organs			
Head	(Shatalin)[6] Pegov[7] (–Oct 52)	Aristov[8] (Oct 52–early 53)	Gromov[9]
Deputy heads		(Shatalin)[6]	
	Gromov[9]	Gromov	
	Dedov[10]	Dedov	
	Yenyutin[11]	Yenyutin (–Jan 54)	
	Skulkov[12] (51–May 52)		
		Churayev[13]	Churayev Storozhev[14]
Administrator of Affairs	Krupin[15]	Krupin	Krupin
Agriculture	Kozlov[16]	Kozlov (–Mar 53)	
Defense Industry	Serbin[17]	Serbin	Serbin
Foreign Cadres Section			
Foreign Policy Commission	(Grigoryan)[18]		(abolished)
International Section		(Stepanov)[3]	
Heavy Industry	(Alekseyev)[19]	(abolished)	(abolished)
Industry-Transport Section			Kuzmin[20]
Light Industry	Pegov[21]	(abolished)	(abolished)
Literature and Art[22]			
Machine Building		(abolished)	(abolished)
Planning-Finance-Trade			
Schools	Zimin[23]		
Science-Education	Zhdanov[24]	(abolished)	(abolished)
Science-Culture Section		Rumyantsev[25]	Rumyantsev
Transport	(Chumachenko)[26]	(abolished)	(abolished)
Special Sector	Poskrebyshev[27]		(abolished)
General Section[28]			Malin
Heads of unnamed sections			
	Baranenkov[29]	Baranenkov	Baranenkov
	Gromov[30]		
	Makarov[31]		
		Chesnokov[32]	
		Yegorov[33]	

[1] N. A. Mikhaylov was identified as having been Agitprop head 1952–53 in the *Bolshaya sovetskaya entsiklopediya* yearbooks for 1962 and 1966. His appointment presumably coincided with his election as CC secretary at the end of the October 1952 congress. Prior to October 1952 he had been first secretary of the Komsomol CC, and in March 1953 he became first secretary of Moscow, losing his post as CC secretary. He may have surrendered his Agitprop post already by February 1953, however (see following item).

[2] V. S. Kruzhkov was identified as deputy head of Agitprop in the 24 February 1950 *Pravda* and as deputy head of an unnamed section in the 30 November 1950 *Vechernyaya Moskva*. On 6 February 1953, *Vechernyaya Moskva* listed him as head of an unnamed section, suggesting that Mikhaylov had already given up the post of Agitprop head. By late 1953 Kruzhkov clearly headed Agitprop: he was identified as head of Agitprop in the 4 December 1953 and 1 February 1954 *Pravda* and in the 15 February 1955 *Sovetskaya kultura*.

[3] V. P. Stepanov was identified as deputy head of Agitprop in the 28 April 1951 *Vechernyaya Moskva*. He had returned to Agitprop following the 1951 closing of *Kultura i zhizn,* of which he was chief editor. His biography in the 1962 yearbook of the *Bolshaya sovetskaya entsiklopediya* lists him in the CC apparat, 1951–55. He was identified as head of an unnamed section during 1954, when he accompanied Khrushchev on visits to Czechoslovakia and China (*Pravda,* 10 June 1954, 29 and 30 September 1954, and 11 and 13 October 1954). Since all these occasions were trips to visit foreign countries, he may briefly have headed the International Section.

[4] In his obituary in the 27 October 1978 *Moskovskaya pravda,* L. A. Slepov was listed as having been deputy head of Agitprop from about 1948 to about 1952, when he became a member of the secretariat of the Ideological Commission under the new CC Presidium. He was identified as editor of *Pravda*'s party life department in the 12 April 1953 *Radyanska Ukraina*.

[5] D. M. Popov, identified as deputy head of Agitprop in the 31 January 1951 *Kultura i zhizn,* was listed as deputy head of Agitprop from 1949 until his death (obituary in the 9 January 1952 *Pravda*).

[6] N. N. Shatalin, identified only as a CC inspector in the 25 February 1950 *Pravda,* was listed as head of an unnamed section in the 30 November 1950 *Vechernyaya Moskva*. As he was former deputy chief of the Cadres Administration, this reference was presumably to the Party, Trade Union, and Komsomol Organs Section. In early 1953, however, he was listed as *deputy* head of an unnamed section (*Vechernyaya Moskva,* 5 February 1953)—suggesting that he had been demoted to make way for appointment of a new cadres chief.

[7] N.M. Pegov apparently became head of the section in 1951 or, at any rate, by mid-1952. His biography in the 1979 *Deputaty verkhovnogo soveta SSSR* lists him as head of the Light Industry Section, then head of the Section for Party, Trade Union, and Komsomol Organs, 1948–52, then CC secretary and candidate member of the CC Presidium from 1952. His last identification as Light Industry Section head was in the 27 January 1951 *Pravda*. He was elected chairman of the Credentials Commission at the 19th Congress, a job usually held by the head of the cadres department. Hence, he apparently held this post at the time of the congress. No one else has ever been identified as cadres chief during 1951 and early 1952. He apparently surrendered the cadres post at the conclusion of the congress, when he was elected a CC secretary and candidate member of the Presidium, and Chelyabinsk First Secretary Aristov was appointed cadres department head.

[8] A. B. Aristov was listed as having been head of the Party, Trade Union, and Komsomol Organs Section for 1952–53 in volume 51 of the second edition of the *Bolshaya sovetskaya entsiklopediya*. His appointment presumably coincided with his election as CC secretary at the October 1952 post-congress plenum, and his removal from the section presumably coincided with his removal as secretary in March 1953.

Prior to his 1952 appointment, he had been first secretary of Chelyabinsk; following his 1953 removal, he was demoted to chairman of the executive committee in far-off Khabarovsk (vol. 51 of the *Bolshaya*).

[9] Ye. I. Gromov was listed as deputy head of an unnamed section in the 30 November 1950 and 6 February 1953 *Vechernyaya Moskva*s. Since he was identified as head of the Party, Trade Union, and Komsomol Organs Section in the 15 January 1954 *Pravda*, his previous posts were presumably in this section also.

[10] A. L. Dedov was identified as deputy head of the Party, Trade Union, and Komsomol Organs Section in the 21 February 1950 *Pravda* and the 5 March 1950 *Sovetskaya Belorussiya*, and as deputy head of an unnamed section in the 6 February 1953 *Vechernyaya Moskva*. In 1955 he was appointed RSFSR minister of state control (see the new RSFSR government listed in the 27 March 1955 *Pravda*).

[11] G. V. Yenyutin was listed as having been inspector in, then deputy head of the Party, Trade Union, and Komsomol Organs Section from 1951 to 1954 in his biography in the 1962 yearbook of the *Bolshaya sovetskaya entsiklopediya*. He had left his previous post as Zaporozhe first secretary in September 1951 to go off for study, according to the 6 September 1951 *Pravda Ukrainy*. In January 1954 he became first secretary of Kamensk *oblast*.

[12] I. P. Skulkov was identified as deputy head of the Party, Trade Union, and Komsomol Organs Section in the 16 December 1951 *Kazakhstanskaya pravda*. His biographies in the 1962 and 1966 yearbooks of the *Bolshaya sovetskaya entsiklopediya* indicate that he had been in the CC apparat during 1951–52. He became first secretary of Ulyanovsk in May 1952 (*Pravda*, 8 May 1952).

[13] V. M. Churayev was listed as having been in the CC apparat starting 1951 in the 1962 and 1966 yearbooks of the *Bolshaya sovetskaya entsiklopediya*. He was identified as deputy head of the Party, Trade Union, and Komsomol Organs Section in the 17 February 1954 *Pravda*.

[14] Ya. V. Storozhev was identified as deputy head of the Party, Trade Union, and Komsomol Organs Section in the 16 January 1954 *Pravda*.

[15] D. V. Krupin was identified as Administrator of Affairs in the 30 November 1950 and 6 February 1953 issues of *Vechernyaya Moskva*.

[16] A. I. Kozlov was listed as head of the Agriculture Section in the 21 February 1950 *Pravda*, 2 December 1950 *Moskovskiy bolshevik*, 20 January 1953 *Vechernyaya Moskva*, and 27 January 1953 *Moskovskaya pravda*. In March 1953 he was appointed minister of agriculture.

[17] I. D. Serbin was listed as head of an unnamed section in the 1 December 1950 *Vechernyaya Moskva* (although his biography in the 1962 yearbook of the *Bolshaya sovetskaya entsiklopediya* says that he became head of a section only in 1958).

[18] V. G. Grigoryan, identified as a "CC worker" in the 23 February 1950 *Pravda*, was listed as head of an unnamed section in the 30 November 1950 *Vechernyaya Moskva*. He had been deputy chief of Agitprop in 1947, but in view of his involvement in foreign affairs by 1950 (he was named a member of the foreign affairs commission of the Council of Nationalities at the 12 June 1950 Supreme Soviet session—the only CC official on this body), his section may have been the Foreign Policy Commission during this period. This presumption is reinforced by Ilya Ehrenburg's mention that Grigoryan handled approval of a 1949 foreign-policy statement. Ehrenburg wrote that when he wrote his speech to deliver at the April 1949 World Peace Congress, he submitted it to the CC for approval and was called in by Grigoryan, "who held a rather high position" and who expressed his approval of it (Ehrenburg, *Post-War Years 1945–1954*, p. 134). In the 5 February 1953 *Vechernyaya Moskva*, he was only listed as a "CC worker," suggesting that he may no longer have headed the section. He was transferred to the foreign ministry in 1953, perhaps in connection with the fall of his patron Beriya.

[19] V. I. Alekseyev was listed as head of an unnamed section in the 1 December 1950 *Vechernyaya Moskva* and as deputy head of an unidentified section in the 7 February 1953 *Vechernyaya Moskva*. Alekseyev presumably headed the Heavy Industry Section, since his whole career was in heavy industry, especially coal. His obituary in the 13 April 1979 *Sotsialisticheskaya industriya* indicated that he had graduated from the Sverdlovsk mining institute, worked in the coal industry in Sverdlovsk and Tula, and had been a "responsible worker of the CC apparat" from 1941 to 1956. In addition, he was credited with making a big contribution to heavy industry and to restoring coal mines and plants after World War II. He appears to have headed some sort of Heavy Industry Section within the Cadres Administration even before the branch sections were set up in 1948. He had been identified as head of a section in the Cadres Administration in a 1943 list of awards for officials in the coal industry (*Vedomosti verkhovnogo soveta SSSR*, 31 October 1943), and was similarly identified at a 1947 Stalino coal conference reported in the 22 June 1947 *Pravda Ukrainy*. He was also included in lists of those receiving awards for successes in the coal industry in the 15 January 1948 *Vedomosti verkhovnogo soveta SSSR*, although he was given no identification on that occasion — a practice often used with CC officials. He probably headed the Heavy Industry Section already in late 1948 when it became an independent section, since he signed the 20 March 1949 *Pravda* obituary of a sector head of the Heavy Industry Section.

[20] I. I. Kuzmin, an assistant to an unnamed deputy premier until 1952, became deputy head, then head of the Industry-Transport Section, later of the Machine Building Section, according to his biography in volume 51 of the *Bolshaya sovetskaya entsiklopediya* (second ed.). In 1957 he became Gosplan chairman.

[21] N. M. Pegov was last identified as head of the Light Industry Section in the 27 January 1951 *Pravda*. Judging by his biography in the 1979 *Deputaty verkhovnogo soveta SSSR* (see footnote 7), he was transferred to head of the Party, Trade Union, and Komsomol Organs Section sometime in 1951 or 1952.

[22] According to the *Istoriya kommunisticheskoy partii sovetskogo soyuza*, vol. 5, bk. 2, p. 221, sections for artistic literature and art, schools, and science and higher educational institutions were created out of Agitprop in December 1950.

[23] P. V. Zimin was identified as head of the Schools Section in the 28 February 1951 *Kultura i zhizn* and in the 12 August 1952 *Pravda*.

[24] Yu. A. Zhdanov was identified as head of the Science and Higher Educational Institutions Section in the 16 October 1951 *Sovetskaya Belorussiya*. He was elected a full CC member at the October 1952 congress, but his status in early 1953 is unclear. He wrote an article in the 16 January 1953 *Pravda* but was identified in no post on this occasion. His section appears to have been reorganized and put under someone else in late 1952 (see following entry).

[25] A. M. Rumyantsev apparently became head of the Science-Culture Section sometime in 1952. His biography in the third edition of the *Bolshaya sovetskaya entsiklopediya* lists him as head of the Science Section starting in 1952, and his biography in the journal *Eko* (no. 2, 1974, p. 42) lists him as head of the Science and Culture Section starting 1952. Aleksandr Nekrich, then working in the History Institute, mentions appealing to "head of the Science Section" Rumyantsev on 21 April 1953 (Nekrich, *Otreshis ot strakha*, pp. 78, 112). The first Soviet press identification of Rumyantsev as head of the section (referred to as the Science and Culture Section) appeared in the 1 February 1954 *Pravda*. Although the years cited in biographies are sometimes a little inaccurate, it is quite possible that Rumyantsev moved to Moscow in 1952, before Stalin's death. At the 19th Congress, he was elected a full CC member and a member of the commission to rewrite the Party Program. This status clearly indicates that he was no longer just director of the Ukrainian Institute of Economics in Kharkov.

Appendix 3

²⁶ G. A. Chumachenko was listed as head of an unnamed section in the 30 November 1950 *Vechernyaya Moskva*. Presumably this was the Transport Section, since he had earlier been Ukrainian CC deputy secretary for transport (*Pravda Ukrainy*, 17 February 1946, 12 December 1947, 13 June 1948) and then deputy minister for cadres for the USSR Railroad Ministry (his 17 June 1948 appointment to this post was announced in the 27 July 1948 *Sobraniye postanovleniy pravitelstva SSSR*). He was again identified as deputy minister of railroads in the 5 February 1953 *Vechernyaya Moskva*. He was elected a CC candidate member at the 1952 congress, but may have still been a department head in view of this status. Another possibility is that the section he is identified as heading may mean the political administration of the railroads, which at least during the war had the rights of a CC section. This view is suggested by the fact that he was identified as chief of the political administration of the Ministry of Railroads in the 16 December 1951 *Komsomolskaya pravda*.

²⁷ A. N. Poskrebyshev was removed as chief of Stalin's secretariat sometime in the winter of 1952–53, according to Alliluyeva (*Twenty Letters to a Friend*, p. 208) and Khrushchev (*Khrushchev Remembers*, pp. 274–75). Poskrebyshev was nevertheless still nominated for election to the Moscow city soviet in early 1953 and was identified as a CC official in accounts of this nomination (*Vechernyaya Moskva*, 6 February 1953). He had been publicly identified as head of the special sector (*Osobyy sektor sekretariata*) when nominated for election to the Moscow city soviet in December 1947 (*Vechernyaya Moskva*, 4 December 1947). The failure to identify him as head of this sector in February 1953 did not, however, necessarily prove the assertions of his demotion, since he had been only vaguely identified as a CC member when nominated for the Moscow city soviet in November 1950 (*Vechernyaya Moskva*, 30 November 1950) and when nominated for the RSFSR Supreme Soviet in November 1951 (*Moskovskaya pravda*, 25 November 1951).

²⁸ By 1954 the special sector, which served among other things as Stalin's private secretariat, had been replaced with a General Section (*Obshchiy otdel*). The biography of its head, V. N. Malin, in the 1962 yearbook of the *Bolshaya sovetskaya entsiklopediya*, stated that he had become head of the section in 1954. For more on this special sector or General Section, see Niels Erik Rosenfeldt's *Knowledge and Power: The Role of Stalin's Secret Chancellery in the Soviet System of Government* (Copenhagen, 1978), and Leonard Schapiro, "The General Department of the CC of the CPSU," in *Survey*, Summer 1975, pp. 53–65.

²⁹ F. I. Baranenkov was listed as head of an unnamed section in the 29 November 1950 and 5 February 1953 *Vechernyaya Moskva*. His obituary in the 14 May 1961 *Pravda* identifies him as having worked in the CC apparat from 1939 to 1955, heading an unnamed CC section for "a number of years." It states that in 1955 he was transferred to chief of the cadres administration of the Ministry of Foreign Affairs.

³⁰ G. P. Gromov was listed as head of an unnamed section in the 30 November 1950 *Vechernyaya Moskva*. He was given no identification when his nomination for the Moscow soviet was reported in the 5 February 1953 *Vechernyaya Moskva*. His obituary in the 20 February 1973 *Komsomolskaya pravda* indicates that he was a CC official from 1950 to 1956. His career before and after this, according to the obituary, was in political work in the armed forces; however, since someone else headed the Main Political Administration of the Armed Forces during these years, Gromov's section is still unknown.

³¹ V. Ye. Makarov was identified as head of an unidentified section in the 30 November 1950 *Vechernyaya Moskva*.

³² D. I. Chesnokov was listed as head of an unnamed section in the 5 February 1953

Vechernyaya Moskva. Since he was coeditor of *Kommunist* and an established ideological leader, this section would presumably be Agitprop, but Mikhaylov or Kruzhkov appeared to head Agitprop at this time.

[33] N. V. Yegorov was listed as head of an unnamed section in the 6 February 1953 *Vechernyaya Moskva.*

Bibliography

Books

Akshinskiy, V. S. *Kliment Yefremovich Voroshilov.* Moscow, 1974.

Alliluyeva, Svetlana. *Twenty Letters to a Friend.* Translated by Priscilla Johnson McMillan. New York: Harper & Row, 1967.

——. *Only One Year.* Translated by Paul Chavchavadze. New York: Harper & Row, 1969.

Armstrong, John A. *The Politics of Totalitarianism.* New York: Random House, 1961.

Avtorkhanov, Abdurakhman. *Stalin and the Soviet Communist Party.* New York: Praeger, 1959.

——. *Zagadka smerti Stalina.* Frankfurt am Main: Possev-Verlag, 1976.

Boffa, Giuseppe. *Inside the Khrushchev Era.* Translated by Carl Marzani. New York: Marzani & Munsell, 1959.

Chakovskiy, Aleksandr. *Blokada.* (Historical novel on the siege of Leningrad, serialized in *Znamya*, nos. 10–12, 1968; 1-3, 1970; 6–8, 1971; 1–4, 1973; 11–12, 1974; and 1–5, 1975.)

——. *Pobeda.* (Historical novel about Stalin at the Potsdam Conference, first part serialized in *Znamya*, nos. 10–12, 1978.)

Conquest, Robert. *Power and Policy in the U.S.S.R.* New York: St. Martin's, 1961.

Dedijer, Vladimir. *Josip Broz Tito: prilozi za biografiju.* Belgrade: Prosveta, 1953.

——. *Tito Speaks.* London: Weidenfeld and Nicholson, 1953.

——. *The Battle Stalin Lost.* New York: Viking, 1971.

Deputaty verkhovnogo soveta SSSR. Moscow, 1962, 1966, 1970, 1974, 1979.

Djilas, Milovan. *Conversations with Stalin.* Translated by Michael B. Petrovich. New York: Harcourt, Brace, and World, Inc., 1962.

Bibliography

Ehrenburg, Ilya. *Post-War Years 1945–1954*. Vol. 6 of *Men, Years—Life*. Translated by Tatiana Shebunina. London: MacGibbon & Kee, 1966.

Fainsod, Merle. *How Russia is Ruled*. Cambridge, Mass.: Harvard University Press, 1964.

Feis, Herbert. *From Trust to Terror: The Onset of the Cold War, 1945–1950*. New York: Norton, 1970.

Graham, Loren. *Science and Philosophy in the Soviet Union*. New York: Alfred A. Knopf, 1972.

Gusarov, Vladimir. *Moy papa ubil Mikhoelsa*. Frankfurt am Main: Possev-Verlag, 1978.

Hoffman, Erik P. and Frederic J. Fleron, Jr. *The Conduct of Soviet Foreign Policy*. Chicago: Aldine, Atherton, 1971.

Hyde, H. Montgomery. *Stalin*. New York: Farrar, Straus & Giroux, 1971.

Istoriya velikoy otechestvennoy voyny sovetskogo soyuza, 1941–1945. 6 vols. Moscow, 1960–63.

Istoriya kommunisticheskoy partii sovetskogo soyuza. Vol. 5, bk. 1. Moscow, 1970. Vol. 5, bk. 2. Moscow, 1980.

Jakobson, Max. *Finnish Neutrality*. London: Hugh Evelyn, 1968.

Joravsky, David. *The Lysenko Affair*. Cambridge, Mass.: Harvard University Press, 1970.

Karasev, A. V. *Leningradtsy v gody blokady 1941–43*. Moscow, 1959.

Katz, Abraham. *The Politics of Economic Reform in the Soviet Union*. New York: Praeger, 1972.

Kedrov, B. M. *Engels i yestestvoznaniye*. Moscow, 1947.

——. *O kolichestvennykh i kachestvennykh izmeneniyakh v prirode*. Moscow, 1947.

Khrushchev, N. S. *The Crimes of the Stalin Era*. Report to the 20th CPSU Congress. (Khrushchev's "Secret Speech.") Reprint from the *New Leader*, 16 July 1956.

——. *Stroitelstvo kommunizma v SSSR i razvitiye selskogo khozyaystva*. 8 vols. Moscow, 1962–64.

——. *Khrushchev Remembers*. Translated by Strobe Talbott. Boston: Little, Brown, 1970.

——. *Khrushchev Remembers: The Last Testament*. Translated by Strobe Talbott. Boston: Little, Brown, 1974.

——. Transcripts of tapes on which the two volumes of *Khrushchev Remembers* are based. Russian Institute, Columbia University.

Kolotov, V. V. and G. A. Petrovichev. *N. A. Voznesenskiy: biograficheskiy ocherk*. Moscow, 1963.

Kolotov, V. V. *Nikolay Alekseyevich Voznesenskiy*. Moscow, 1974, 1976.

KPSS v rezolyutsiyakh i resheniyakh syezdov, konferentsiy i plenumov TsK. Pt. 3, 7th ed., Moscow, 1954; and vol. 6, 8th ed., Moscow, 1971.

Leningrad: entsiklopedicheskiy spravochnik. Moscow and Leningrad, 1957.

Mastny, Vojtech. *Russia's Road to the Cold War: Diplomacy, Warfare, and the Politics of Communism, 1941-1945.* New York: Columbia University Press, 1979.

McCagg, William O., Jr. *Stalin Embattled, 1943-1948.* Detroit: Wayne State University Press, 1978.

Medvedev, Roy. *Let History Judge.* Edited by David Joravsky and Georges Haupt. Translated by Colleen Taylor. New York: Alfred A. Knopf, 1971.

Medvedev, Zhores A. *The Rise and Fall of T. D. Lysenko.* Collected and translated by I. M. Lerner. New York: Columbia University Press, 1969.

Narodnoye khozyaystvo SSSR v 1958 g. Moscow, 1958.

Nekrich, Aleksandr. *Otreshis ot strakha.* London: Overseas Publications Interchange Ltd., 1979.

Nicolaevsky, Boris I. *Power and the Soviet Elite.* Edited by Janet D. Zagoria. New York: Praeger, 1965.

Paterson, Thomas G. *Soviet-American Confrontation: Postwar Reconstruction and the Origins of the Cold War.* Baltimore: The Johns Hopkins University Press, 1973.

Patolichev, N. S. *Ispytaniye na zrelost.* Moscow, 1977.

Petrov, Yuriy P. *Partiynoye stroitelstvo v sovetskoy armii i flote (1918-1961).* Moscow: Voyennoye izdatelstvo, 1964.

——. *Stroitelstvo politorganov, partiynykh i komsomolskikh organizatsiy armii i flota (1918-1968).* Moscow: Voyennoye izdatelstvo, 1968.

Ploss, Sidney. *Conflict and Decision-Making in Soviet Russia.* Princeton, N.J.: Princeton University Press, 1965.

Politicheskiy dnevnik. Vol. 1. Amsterdam: Alexander Herzen Foundation, 1972.

Politicheskiy dnevnik. II. Amsterdam: Alexander Herzen Foundation, 1975.

Resheniya partii i pravitelstva po khozyaystvennym voprosam. Vol. 3, 1941-52. Moscow, 1968.

Rodionov, P. A. *Kollektivnost: vysshiy printsip partiynogo rukovodstva.* Moscow, 1974.

Rosenfeldt, Niels Erik. *Knowledge and Power: The Role of Stalin's Secret Chancellery in the Soviet System of Government.* Kobenhavns Universitets Slaviske Institut, Study No. 5. Copenhagen: Rosenkilde and Bagger, 1978.

Salisbury, Harrison E. *The 900 Days.* New York: Harper & Row, 1969.

Schapiro, Leonard. *The Communist Party of the Soviet Union.* New York: Random House, 1959.

Solzhenitsyn, Aleksandr. *Bodalsya telenok s dubom.* Paris: YMCA Press, 1975. (Published in English as *The Oak and the Calf.* Translated by Harry Willetts. New York: Harper & Row, 1979.)

Stalin, I. V. *Sochineniya.* Vol. 3 (XVI), 1946-53. Ed. Robert H. McNeal. Stanford, Hoover Institution on War, Revolution and Peace, Stanford

Bibliography

University, 1967. (Stalin's postwar statements, in Russian).

Stenograficheskiy otchet, XVIII syezd vsesoyuznoy kommunisticheskoy partii (b). Moscow, 1939.

Sulzberger, C. L. *The Big Thaw.* New York: Harper & Row, 1956.

Swayze, Harold. *Political Control of Literature in the USSR, 1946–59.* Cambridge, Mass.: Harvard University Press, 1962.

The Soviet-Yugoslav Dispute. London: Royal Institute of International Affairs, 1948.

Tikos, Laszlo. E. *Vargas Taetigkeit als Wirtschaftsanalytiker und Publizist.* Cologne: Boehlau-Verlag, 1965.

(Varga, Ye. S.) *Soviet Views on the Post-War World Economy.* (An Official Critique of Eugene Varga's *Changes in the Economy of Capitalism Resulting from the Second World War,* translated by Leo Gruliow). Washington D. C.: Public Affairs Press, 1948.

Vloyantes, John P. *Silk Glove Hegemony: Finnish-Soviet Relations 1944–1974.* The Kent State University Press, 1975. (No city listed)

Voslensky, Michael S. *Nomenklatura: Die herrschende Klasse der Sowjetunion.* Vienna: Verlag Fritz Molden, 1980.

Voznesenskiy, N. A. *Voyennaya ekonomika SSSR v period otechestvennoy voyny.* Moscow, 1947.

Werth, Alexander. *Russia: The Post-War Years.* London: Robert Hale & Co., 1971.

Yergin, Daniel. *Shattered Peace: The Origins of the Cold War and the National Security State.* Boston: Houghton Mifflin, 1977.

Zemtsov, Ilya. *Partiya ili mafiya?* Paris: Les Editeurs Reunis, 1976.

Zinovyev, Aleksandr. *Svetloye budushcheye.* Lausanne: Editions l'Age d'Homme, 1978. (Published in English as *The Radiant Future.* New York: Random House, 1981.)

Selected Articles

Abramov, B. A., "Organizatsionno-partiynaya rabota KPSS v gody chetvertoy pyatiletki," *Voprosy istorii KPSS,* March 1979, pp. 55–65.

Barghoorn, Frederick C., "The Varga Discussion and Its Significance," *The American Slavic and East European Review,* no. 3 (1948), pp. 214–36.

Berg, Raisa, "Povest o genetike," *Vremya i my* (Tel Aviv), no. 50 (1980), pp. 162–202.

Bialer, Seweryn, "I Chose Truth," *News from Behind the Iron Curtain,* October 1956.

Cohen, Stephen F., "The Friends and Foes of Change: Reformism and Conservatism in the Soviet Union," *Slavic Review,* vol. 38, no. 2 (June 1979), pp. 187–202.

Ducoli, John, "The Georgian Purges (1951–53)," *Caucasian Review,* no. 6 (1958), pp. 54–61.

Harris, Jonathan, "The Origins of the Conflict Between Malenkov and Zhdanov: 1939–1941," *Slavic Review,* vol. 35, no. 2 (June 1976), pp. 287–303.

Kennan, George F., "Excerpts from a Draft Letter Written at Some Time During the First Months of 1945," *Slavic Review,* vol. 27, no. 3 (September 1968), pp. 481–84.

Kolotov, V. V., "Ustremlennyy v budushcheye," *Znamya,* June 1974.

Krotkov, Yu., "Konets Marshala Beriya," *Novyy zhurnal,* December 1978, pp. 212–32.

Maleyko, L. A., "Iz istorii razvitiya apparata partiynykh organov," *Voprosy istorii KPSS,* February 1976, pp. 111–122.

Medvedev, Zhores A., "Biologicheskaya nauka i kult lichnosti," *Grani,* nos. 70 and 71 (February and May 1969).

Nekrich, Alexander, "The Arrest and Trial of I. M. Maysky," *Survey,* nos. 100–01 (Summer-Autumn 1976), pp. 313–20.

——, "'The Menshevistic Idealist': Abram Deborin," *Survey,* no. 105 (Autumn 1977–78), pp. 137–43.

Nemzer, Louis, "The Kremlin's Professional Staff: The 'Apparatus' of the Central Committee, Communist Party of the Soviet Union," *American Political Science Review,* vol. 44, no. 1 (March 1950), pp. 64–85.

Nicolaevsky, Boris, "Palace Revolution in the Kremlin," *New Leader,* 19 March 1949.

——, "Malenkov—his Rise and his Policy," *New Leader,* 23 March 1953.

——, "How Malenkov Triumphed," *New Leader,* 30 March 1953.

——, "How Did Stalin Die?," *New Leader,* 20 April 1953.

——, "The Abakumov Case," *New Leader,* 10 January 1955.

Nikolayevskiy, Boris, "Na komandnykh vysotakh Kremlya," *Sotsialisticheskiy vestnik,* 22 May 1946, and 21 June 1946.

Nordahl, Richard, "Stalinist Ideology: The Case of the Stalinist Interpretation of Monopoly Capitalist Politics," *Soviet Studies,* vol. 26, no. 2 (April 1974), pp. 239–59.

Polozov, G. P., "Partiynoye rukovodstvo deyatelnostyu intelligentsii v gody velikoy otechestvennoy voyny," *Voprosy istorii KPSS,* September 1976, pp. 116–25.

Schapiro, Leonard, "The General Department of the CC of the CPSU," *Survey,* no. 96 (Summer 1975), pp. 53–65.

Encyclopedias

Bolshaya sovetskaya entsiklopediya [Great Soviet encyclopedia], 2d and 3d editions.

Ekonomicheskaya entsiklopediya [Economic encyclopedia].

Filosofskaya entsiklopediya [Philosophical encyclopedia], vol. 1, 1960; vol. 3, 1964; vol. 5, 1970.

Bibliography

Malaya sovetskaya entsiklopediya [Small Soviet encyclopedia].
Sovetskaya istoricheskaya entsiklopediya [Soviet historical encyclopedia].
Sovetskaya voyennaya entsiklopediya [Soviet military encyclopedia].
Ukrainska radyanska entsiklopediya [Ukrainian Soviet encyclopedia].

Soviet Newspapers

Bakinskiy rabochiy, organ of the Azerbaydzhan CC.
Izvestiya, organ of the USSR Supreme Soviet Presidium.
Kazakhstanskaya pravda, organ of the Kazakhstan CC.
Kommunist, organ of the Armenian CC.
Kommunist Tadzhikistana, organ of the Tadzhik CC.
Komsomolskaya pravda, organ of the Komsomol CC.
Kultura i zhizn, organ of Agitprop (1946–51).
Leningradskaya pravda, organ of the Leningrad party organization.
Literaturnaya gazeta, organ of the USSR Writers' Union.
Moskovskaya pravda, organ of the Moscow party organization (earlier named *Moskovskiy bolshevik*).
Moskovskiy bolshevik, organ of the Moscow party organization (later renamed *Moskovskaya pravda*).
Pravda, main organ of the CC.
Pravda Ukrainy, Russian-language organ of the Ukrainian CC.
Pravda vostoka, organ of the Uzbek CC.
Radyanska Ukraina, Ukrainian-language organ of the Ukrainian CC.
Selskaya zhizn, CC newspaper for agriculture.
Sotsialisticheskaya industriya, CC newspaper for industry.
Sovetskaya Belorussiya, organ of the Belorussian CC.
Sovetskaya Estoniya, organ of the Estonian CC.
Sovetskaya Kirgiziya, organ of the Kirgiz CC.
Sovetskaya Latviya, organ of the Latvian CC.
Sovetskaya Litva, organ of the Lithuanian CC.
Sovetskaya Rossiya, CC newspaper for the RSFSR.
Turkmenskaya iskra, organ of the Turkmen CC.
Uchitelskaya gazeta, organ of the USSR Ministry of Education.
Vechernyaya Moskva, organ of the Moscow party organization.
Zarya vostoka, organ of the Georgian CC.

Soviet Journals

Bolshevik, leading theoretical and political journal of the CC (later renamed *Kommunist*).
Eko (*Ekonomika i organizatsiya promyshlennogo proizvodstva*), journal of the Siberian division of the USSR Academy of Sciences.

230

Filosofskiye nauki, organ of the Ministry of Higher and Secondary Specialized Education.

Izvestiya akademii nauk SSSR, seriya ekonomicheskaya, economic journal of the USSR Academy of Sciences.

Izvestiya akademii nauk SSSR, seriya istorii i filosofii, historical and philosophical journal of the USSR Academy of Sciences.

Kommunist, leading theoretical and political journal of the CC (earlier named *Bolshevik*).

Krokodil, satire magazine.

Mirovoye khozyaystvo i mirovaya politika, journal of the Institute for World Economics and World Politics.

Molodoy kommunist, journal of the Komsomol CC.

Moskva, journal of the RSFSR and Moscow writers' unions.

Novaya i noveyshaya istoriya, current history journal of the General History Institute of the Academy of Sciences.

Novyy mir, journal of the USSR Writers' Union.

Oktyabr, journal of the RSFSR Writers' Union.

Partiynaya zhizn, CC journal for party organizational work.

Partiynoye stroitelstvo, CC journal for party organizational work (replaced in 1946 by *Partiynaya zhizn*).

Perets, Ukrainian satire magazine.

Planovoye khozyaystvo, organ of Gosplan.

Pod znamenem marksizma, "philosophical and socioeconomic journal" (not specifically identified with any organization).

Problemy mira i sotsializma, organ of the international communist movement issued in several languages in Prague.

Sobraniye postanovleniy pravitelstva SSSR, bulletin of the Council of Ministers.

Vedomosti verkhovnogo soveta SSSR, bulletin of the Supreme Soviet.

Vestnik akademii nauk SSSR, journal of the USSR Academy of Sciences.

Voprosy ekonomiki, journal of the Institute of Economics.

Voprosy filosofii, journal of the Institute of Philosophy.

Voprosy istorii, journal of the Academy of Sciences' history division and the USSR Ministry of Higher and Secondary Specialized Education.

Voprosy istorii KPSS, journal of the Institute of Marxism-Leninism.

Zhurnalist, journal of the Journalists' Union.

Znamya, journal of the USSR Writers' Union.

Zvezda, journal of the USSR Writers' Union, published in Leningrad.

Index

Index

Aristov, A. B., 140, 140n, 141, 141n, 145, 145n, 146, 146n, 147, 149, 190, 191, 218, 219, 220
Arkhiptsev, F. T., 178
Artistic Literature and Art Section, of CC, 206, 209, 218, 221
Arutyunov (Arutinov), G. A., 139, 139n
Arzhanov, M. A., 86
Avakyan, A. A., 80
Avtorkhanov, Abdurakhman, 87n

Bagirov, M. D., 50, 50n, 55, 139, 191, 192
Bailey, 25n
Bakradze, V. M., 143n
Baranenkov, F. I., 218, 222
Baskin, M. P., 73
Belous, S. K., 134
Belov, P. A., 152
Berezkin, V. A., 61
Berg, Raisa, 118n
Beriya, L. P., 20, 22, 26, 35, 37, 39n, 40, 41, 43n, 44, 45n, 50, 51n, 52n, 55, 69, 77, 77n, 94, 96, 98, 103, 103n, 104, 105, 105n, 106, 106n, 107, 107n, 112, 113, 123n, 126, 127, 131n, 133, 133n, 134, 135, 136, 137, 139, 141, 142, 143, 143n, 145, 146, 146n, 147, 148, 148n, 149, 154n, 156, 156n, 157, 158, 160, 163, 164, 184, 188–97, 220
  involvement in Leningrad Case, 112, 114, 122, 123n, 124, 125, 130n, 133, 184
Bialer, Seweryn, 124n
Bikkenin, N. B., 180
Biology debate, 13, 14, 67, 68, 69, 78–84, 94, 162
Boffa, Giuseppe, 124, 124n, 158n
Bohr, Niels, 82
Bokshitskiy, M. L., 89, 91, 92
Bolshevik, 26, 29, 29n, 76, 155
  changes in, 26, 29, 29n, 30, 77n, 111, 153, 160n
Borkov, G. A., 46, 53, 54, 61n, 108, 213, 214
Borshchagovskiy, Aleksandr, 119, 120n
Brezhnev, L. I., 33, 147, 149, 176, 190, 191
Bulganin, N. A., 37, 37n, 40, 41, 43, 45n, 103n, 145, 147, 148, 188, 189, 190, 193, 194, 195, 196, 197
Bureaus of CC (*see* Central Committee bureaus)

Burmeister, Alfred, 201n
Bushinskiy, B. P., 115
Byrnes, James, 24n

Cadres Administration, of CC, 42, 43, 43n, 44, 46, 47, 47n, 50, 52, 52n, 53, 53n, 61n, 62, 65, 108, 108n, 109, 125n, 135n, 140, 198, 199, 200, 200n, 201, 204, 205, 205n, 209, 213, 214, 217, 219, 221
Central Committee apparat, 34, 44, 47
  reorganization of, 14, 20, 40, 43, 47, 48, 51, 52, 94, 104, 108, 108n, 109, 184, 198–209
Central Committee bureaus
  for Estonia, 202
  for Far East, 127
  for Latvia, 202
  for Lithuania, 43, 109, 202
  for Moldavia, 202
Central Committee decrees, 29, 29n, 30, 30n, 31n, 45, 49, 50, 51, 52, 54, 55, 56n, 57, 58, 59, 60n, 61, 61n, 63, 64n, 72, 73, 73n, 77n, 102, 108, 110, 110n, 111, 111n, 113, 125n, 128, 128n, 133n, 139n, 143, 155, 159n, 199, 211
Central Committee inspectors, establishment of post of, 46, 47, 54, 55, 203
Central Committee plenums, 33, 39, 40, 40n, 41n, 43, 44, 52n, 60n, 62n, 63, 63n, 64, 65, 90n, 104, 124, 130, 138n, 140n, 144, 144n, 145n, 148, 158n, 159, 184, 188, 189, 191, 192, 214, 219
Central Committee sections (*see* title of individual section)
Chakovskiy, Aleksandr, 38, 201
Charkviani, K. N., 143
Checking Administration, of CC, 43n, 44, 46, 47, 52n, 53, 54, 55, 56, 59, 61n, 65, 108, 142, 203, 204, 213, 214, 217
Cheremnykh, P. S., 111
Chernyak, Ya., 120
Chertkov, V. P., 165
Chervonenko, S. V., 208
Chesnokov, D. I., 76, 76n, 78n, 83, 122, 122n, 153, 154, 154n, 160, 160n, 190, 218, 222, 223
Chugayev, A. Ya., 154n

Index

Index

Index

Index

Library of Congress Cataloging in Publication Data

Hahn, Werner G.
  Postwar Soviet politics.

  Bibliography: p.
  Includes index.
  1. Soviet Union—Politics and government—1936–1953. 2. Zhdanov, Andrei
Aleksandrovich, 1896–1948. 3. Learning and scholarship—Soviet Union. 4. Science
and state—Soviet Union. I. Title.
DK267.H23   947.084′2   81-15234
ISBN 0-8014-1410-5   AACR2